Tetyana Shapoval

Local imaging of magnetic flux in superconducting thin films

Tetyana Shapoval

Local imaging of magnetic flux in superconducting thin films

From pinning to depinnig of vortices

Südwestdeutscher Verlag für Hochschulschriften

Imprint
Any brand names and product names mentioned in this book are subject to trademark, brand or patent protection and are trademarks or registered trademarks of their respective holders. The use of brand names, product names, common names, trade names, product descriptions etc. even without a particular marking in this work is in no way to be construed to mean that such names may be regarded as unrestricted in respect of trademark and brand protection legislation and could thus be used by anyone.

Publisher:
Südwestdeutscher Verlag für Hochschulschriften
is a trademark of
Dodo Books Indian Ocean Ltd., member of the OmniScriptum S.R.L Publishing group
str. A.Russo 15, of. 61, Chisinau-2068, Republic of Moldova Europe
Printed at: see last page
ISBN: 978-3-8381-2475-9

Zugl. / Approved by: Dresden, TU, Diss., 2010

Copyright © Tetyana Shapoval
Copyright © 2011 Dodo Books Indian Ocean Ltd., member of the OmniScriptum S.R.L Publishing group

Contents

1	**Introduction**	**1**
2	**Superconductors in magnetic fields**	**5**
	2.1 Vortex state	6
	2.2 The low field limit	8
	2.3 Flux line pinning	11
	2.4 Bean's model of the critical state	12
3	**Local vortex imaging techniques**	**15**
	3.1 Bitter decoration	16
	3.2 Scanning tunnelling microscopy and spectroscopy	17
	3.3 Magnetic force microscopy	18
	3.4 Scanning Hall probe microscopy	19
	3.5 Magneto-optics	19
	3.6 Lorentz microscopy	20
	3.7 Scanning SQUID microscopy	21
4	**Experimental methods**	**23**
	4.1 Low temperature magnetic force microscopy	23
	4.2 Other techniques	28
5	**Vortices and defects in $YBa_2Cu_3O_{7-\delta}$ thin films**	**29**
	5.1 Surface roughness and growth process	30
	5.2 Vortex imaging on flat off-axis PLD films	31
	5.3 Local and global measurements of the critical current	34
	5.4 Artificial defects	38
	5.5 Is there any possibility to correlate vortices with defects?	41
6	**Nanoscale wedge polishing of thin films**	**45**
	6.1 Experimental details	46
	6.2 Plane polishing of YBCO films	48
	6.3 Wedge polishing of nano-engineered YBCO films	49
	6.4 Other applications	53

Contents

7	**Pinning investigation in NbN thin films**	**55**
	7.1 Tip-vortex interaction: monopole model	55
	7.2 Local depinning of individual flux lines	57
	7.3 Global estimation of the pinning force	59
8	**Nb/Py hybrid structure: pinning by magnetic dots**	**63**
	8.1 Array of Permalloy dots	64
	8.2 Low temperature experiments and discussion of the pinning mechanism	67
9	**Conclusions**	**73**
	List of Publications	**77**
	Bibliography	**79**
	List of Figures	**93**
	List of Tables	**99**
	Acknowledgments	**101**

1 Introduction

Once upon a time in Russia... It was the year 1950 when the great theoreticians Landau and Ginzburg published their thermodynamic (GL) theory of superconductivity [Gin50] and introduced the dimensionless Ginzburg-Landau parameter κ.

The critical field calculated from GL theory perfectly fits the experimental data for at that time existing pure classical superconductors with $\kappa \ll 1$ (type I superconductors).

At the same time a PhD student of Landau, Alexey Abrikosov, who curiously followed the question: "What will happen in the opposite case?", predicted an absolutely new kind of behavior of magnetic field in the superconductor with $\kappa > \sqrt{\frac{1}{2}}$ [Abr85]. His famous paper published in 1957 describes the possibility of magnetic flux lines to penetrate the superconducting material in a regular array of flux quanta now often called Abrikosov vortices [Abr57]. This was the beginning of the era of type II superconductors, that not only brought a new exciting phenomenon to the scientific world, but also opened the possibility of high-field applications of superconducting materials.

In absolutely clean "ideal" superconductors vortices can easily be moved by applying a current under the influence of a Lorentz force. This leads to non-zero resistivity immediately after the current has been applied. But, as nothing is perfect in reality, all materials contain defects. It was found that defects with a size comparable to the GL coherence length can strongly enhance the critical current density [Tin96].

The application of superconducting cables for the construction of high field magnets is impossible without pinning, i.e. the property of a superconducting vortex to be localized at the position with minimum energy (on defects and inhomogeneities) and being fixed (pinned) there during the application of a current. Thus, current without any losses can be transported by these materials until the driving force induced by the applied current or magnetic field will exceed the pinning force of the defects.

When Bednorz and Müller discovered high temperature superconductors (HTSC) in 1986 [Bed86], the superconducting community experienced a second break-through. From the moment when T_c exceeded the boiling temperature of liquid nitrogen [Wu87] a wide industrial application of superconductors came much closer to reality. Until now the most promising

1 Introduction

candidates for high power applications are $YBa_2Cu_3O_{7-\delta}$ (YBCO) coated conductors. Because of the growth mechanism, oxygen vacancies and the low coherence length, this material has already a high percentage of natural defects, thus, quite strong pinning. The great attention of the scientific community concentrates on the improvement of the critical current density (J_c) of YBCO coated conductors by implication of various defects with high pinning potential, trying to reach the value needed to be economically competitive with standard copper wires. Despite the high amount of reported studies that show a strong enhancement of J_c in the presence of various artificial nanodefects, a clear microscopical understanding of the pinning mechanism is still missing.

Low temperature type-II superconductors (LTSC) are usually preferable candidates for basic studies of the vortex matter in thin films. Due to their rather high homogeneity, these films allow to probe the effect of the artificially modified pinning landscape on the vortex arrangement. Therefore, during the last decades artificial defects have been incorporated into a superconducting matrix in different manners: random, periodical, holes, inclusions and, recently, magnetic materials causing a variety of exciting phenomena. One of them is an interplay of superconductivity and magnetism. In ferromagnetic/superconducting (FM/SC) hybrid structures it was demonstrated that the vicinity of FM materials, formerly expected to be destructive for the superconductivity, can even lead to a field-induced superconductivity and strongly enhanced pinning of vortices (see e.g. [Lan03, Hof08]). This counterintuitive interplay of magnetism and superconductivity is today a topic of high scientific interest with many phenomena needed to be investigated and explained.

The first local visualization of the Abrikosov vortex lattice was performed by Bitter decoration on lead single crystals in 1967 [Ess67] followed by a variety of experiments on local imaging of the vortex lattice. After the invention of the scanning tunneling microscope (STM) in 1981 and the atomic force microscope (AFM) in 1985, the latter also used as magnetic force microscope MFM, a large amount of local studies were performed first on single crystals and then also on thin films with a high surface quality. Later, magneto-optics, scanning Hall probe microscopy and Lorenz microscopy found their niche in the vortex visualization.

The following work concentrates on the well known and widely studied problem of pinning mechanisms of vortices in superconducting thin films. This problem is still very up-to-date, since the understanding of this phenomenon can give a key to tune critical parameters of the superconducting materials making them useful for different applications. The main aspects of this thesis are: (i) the vortex visualization on thin films with different surface quality (from nanometer smooth Nb films to rough industrially applicable YBCO films); (ii) the possibility to correlate vortices with artificial defects; and (iii) the depinning of individual vortices from the natural and artificial defects that allows to probe the local modulation of the pinning force.

Starting from the theory of type II superconductors (Chapter 2), with the main focus on their behavior in magnetic fields, this work will guide the reader through different methods of local imaging of vortices (Chapter 3), showing their advantages and disadvantages for the chosen materials.

Chapter 4 is devoted mostly to the description of the low temperature magnetic force microscope (LT-MFM) used here for vortex imaging (Omicron Cryogenic SFM) briefly touching all other techniques used in this work.

The results are presented in four chapters, which are separated into two main parts. In the first part, Chapter 5 describes the work done within the framework of the European project HIPERCHEM. The goal of the project was to increase the J_c of coated conductors to an industrially applicable value by introducing various artificial defects. The own contribution was to establish a correlation between the locations of superconducting vortices and artificial defects in YBCO thin films by local imaging and, in that way, to estimate the pinning potential of different defects for a better microscopical understanding of the pinning mechanism. Chapter 5 also discusses all problems and challenges that appeared on the way to the vortex imaging on very rough but industrially attractive thin films. One of them is the surface roughness (droplets, precipitates) that causes a severe problem to the scanning MFM tip. To overcome this problem a unique nanoscale wedge polishing technique was developed. This method, presented in Chapter 6, provides a possibility to perform vortex imaging on rough as-prepared samples and opens an easy way to look into the interior of thin films by different surface sensitive techniques.

The second part of the results is devoted mostly to the basic studies of the pinning mechanism in LTSC thin films. In Chapter 7 the vortex lattice modification by natural pinning is visualized in NbN thin films. The temperature dependence of the vortex distribution as well as depinning of vortices by the MFM tip are reported. The pinning force estimated locally from the tip-vortex interaction based on the monopole-monopole model is compared with the global value calculated from the transport measurements. In Chapter 8 the tip-vortex interaction force is used to probe the strongly enhanced pinning of vortices by an ordered array of FM dots in a FM/SC hybrid structure (Py = $Ni_{80}Fe_{20}$/Nb).

As this thesis is dealing with the characterization of thin films, the preparation methods for the respective samples are briefly mentioned in each chapter supported with references to the corresponding literature. The main results are summarized in the conclusions.

2 Superconductors in magnetic fields

> Vortex is a whirl of activities.
> E. H. Brandt et al. [Bra02b]

This chapter briefly summarizes the main theoretical aspects that describe the behavior of superconducting materials in a magnetic field. The term *superconductivity* appeared in the year 1911, when Heike Kammerlingh Onnes unexpectedly and unpredictably found the sudden drop of the electrical resistance of mercury at 4.2 K, the boiling temperature of liquid helium [Onn11]. The temperature at which the resistance drops to zero is called critical temperature, T_c. Below this temperature the material is a *perfect conductor*, thus a current can pass through it without any dissipation ($R = 0$). More than 20 years later a second phenomenon of this at that time extraordinary class of materials was revealed: *perfect diamagnetism* known as the Meissner-Ochsenfeld effect [Mei33]. Meissner and Ochsenfeld found that an applied magnetic field is expelled from the sample when cooled below T_c, if it does not exceed a certain critical value H_c. It took more than 50 years to get a theoretical explanation of these amazing phenomena. Two equations proposed by the brothers Fritz and Heinz London in 1935 were the first attempt to describe the electromagnetic behavior of superconductors phenomenologically [Lon35]. A thermodynamical approach was made in 1950 by Ginzburg and Landau [Gin50]. The Ginzburg-Landau (GL) theory for superconductors is based on the theory of second-order phase transitions and gives a good macroscopic description of the superconducting state by being still phenomenological. Using this approach, the existence of a new type of superconductors was predicted by Abrikosov [Abr57], finally leading from the pure phenomenon of superconductivity to real industrial applications. And only in 1957 Bardeen, Cooper and Schrieffer have performed a breakthrough in the theoretical understanding of superconductivity: the BCS theory [Bar57] describes the superconducting state by pairing of electrons occupying quantum states with opposite momentum and spins (Cooper pair). Although the BCS theory was able to explain conventional superconductivity, the theoretical understanding of high-temperature superconductivity (HTS) [Bed86] is still missing. The discovery of superconductivity in Re(O,F)FeAs pnictides in February 2008 [Kam08] newly motivates the active search for the better theoretical explanation of this phenomenon bringing

2 Superconductors in magnetic fields

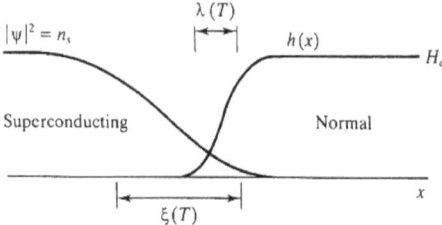

Fig. 2.1: Spatial variation of the microscopic magnetic field $h(x)$ and of the local density of superconducting electrons n_s at the interface between superconducting and normal state [Tin96].

a hope that the mystery of superconductivity can be exposed up to its 100th anniversary.

2.1 Vortex state

The complete microscopic BCS [Bar57] theory gives an excellent explanation of many experimental data in conventional superconductors, but becomes complicated in cases when the spatial inhomogeneities are the objects of interest. In that case, the GL theory is a powerful tool to describe the macroscopic behavior of superconductors such as e.g. the vortex state (mixed state). The GL theory, as it was shown by Gor'kov [Gor59], can be derived from the more general BCS theory and is restricted in its application to small and slow spatial variations of the pseudowavefunction $\psi(r)$ and to the temperature region $T \approx T_c$. The pseudowavefunction is introduced as $|\psi(r)|^2 = n_s(r)$, where n_s is the local density of superconducting electrons and, hence, is equal to zero in the normal state. By minimizing the overall free energy functional, two differential GL equations can be derived [Tin96]. These equations introduce two important characteristic lengths: (i) the *GL coherence length* ξ characterizing the scale of the spatial variation of the $\psi(r)$ function and (ii) the *penetration depth* λ showing how rapidly the magnetic field decays within a superconductor (Fig. 2.1). Both of them exhibit a similar temperature dependence: $\lambda(T) \approx \lambda(0)(1 - \frac{T}{T_c})^{-1/2}$ and $\xi(T) \approx \xi(0)(1 - \frac{T}{T_c})^{-1/2}$ [Bra02a, Tin96]. Hence the ratio $\kappa = \lambda/\xi$, the so called GL parameter, is a material specific dimensionless constant. This parameter allows to distinguish between type-I and type-II superconductors which exhibit a different physical behavior in magnetic fields. For $\kappa < 1/\sqrt{2}$, the superconducting/normal state (SC/NS) phase boundary energy has a positive contribution to the total energy resulting in type-I superconductivity. Such materials expel an applied field from the interior if the field does not exceed the thermodynamic critical field H_c (Meissner phase). The field penetrates only into a thin surface layer which is determined by λ (Fig. 2.1). For larger

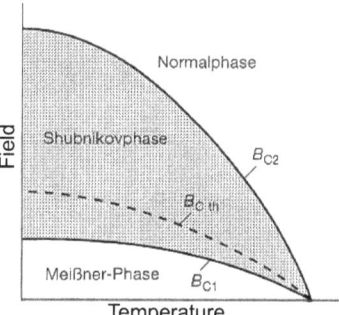

Fig. 2.2: Sketch of the phase diagram of type-II superconductors [Buc04].

fields than H_c the sample switches to the normal state and is no more superconducting. For $\kappa \geq 1/\sqrt{2}$, the SC/NS phase boundary energy is negative. Thus, the superconductor prefers the formation of as many domains as possible to increase the SC/NS surface area [Bra09b]. A. Abrikosov [Abr57] introduced a solution of the GL equations for $\kappa > 1/\sqrt{2}$: a stable periodic solution where the applied field penetrates the superconductor in the form of an ordered lattice of flux lines (vortices). The theoretical prediction of Abrikosov was later confirmed by direct imaging of separated vortices ordered in a hexagonal lattice by Essmann and Träuble [Ess67] (see also Chapter 3).

Type-II superconductors have two characteristic fields: H_{c1} and H_{c2} (Fig. 2.2). Below the first critical field, $B_{c1} = \mu_0 H_{c1} \approx \phi_0 \ln(\sqrt{2}\kappa)/(4\pi\lambda^2)$, with $\mu_0 = 4\pi \times 10^{-7}\,\mathrm{TA/m}$ being the magnetic permeability of vacuum, the material behaves similar to type-I superconductors expelling the field from the interior [Bra02a, Tin96]. When the applied field exceeds H_{c1}, a type-II superconductor does not enter the normal state, but exhibits a mixed state with an Abrikosov lattice of superconducting vortices, also called Shubnikov phase. Each vortex carries one quantum of the magnetic flux:

$$\phi_0 = \frac{h}{2e} = 2.07 \times 10^{-15}\,\mathrm{Tm}^2, \qquad (2.1)$$

which is generated by the supercurrent circulating around the vortex core. The vortex core is defined as an area where the local density of superconducting electrons n_s vanishes. The core radius is a bit larger than ξ [Tin96] and defines the area where the sample is in the normal state. The magnetic field has a maximum in the middle of the vortex and decays exponentially over the distance of λ. The flux quantum ϕ_0 is constant for all materials, temperatures and applied fields. By increasing the applied field H only the density of vortices increases

2 Superconductors in magnetic fields

and hence the lateral distance between them decreases. For a triangular array, which is in most cases the energetically favored lattice, the distance between next neighboring vortices is $a = \left(\frac{4}{3}\right)^{1/4}\sqrt{\frac{\phi_0}{\mu_0 H}} = 1.075\sqrt{\frac{\phi_0}{\mu_0 H}}$ [Tin96].

For low applied fields $H_{c1} \leq H \ll H_{c2}$ the vortices can be considered as separated magnetic objects. Increasing the field, firstly the magnetic flux lines start to overlap and then the vortex cores get closer to each other (Fig. 2.3 (a)). At $H \rightarrow H_{c2}$ the order parameter decreases and fully vanishes when the upper critical field $B_{c2} = \mu_0 H_{c2} = \phi_0/(2\pi\xi^2)$ is achieved [Bra02a]. Above H_{c2} the sample is normal conducting as is sketched in the phase diagram (Fig. 2.2). In HTS H_{c2} is much higher than the thermodynamical critical field $H_{c,th}$, which is rather a theoretical quantity than an experimental one [Tin96]. Consequently, the existence of the mixed state significantly increases the in-field superconductivity and hence opens the unique possibility for industrial applications of superconducting materials to create high fields.

2.2 The low field limit

For the case of a low magnetic induction $B \ll B_{c2}$ and a large $\kappa \gg 1$ the London theory [Lon35] can be applied to calculate the vortex lattice properties. In this limit the total magnetic induction is a linear superposition of the fields coming out from the individual vortices [Bra02a]. Since the low field limit is mostly used in the following thesis, the theory of the separated vortices is described in detail.

The two London equations:

$$\mathbf{E} = \mu_0 \lambda^2 \frac{\partial \mathbf{J}}{\partial t}, \qquad (2.2)$$

$$\mathbf{B} = -\mu_0 \lambda^2 \nabla \times \mathbf{J}, \qquad (2.3)$$

describe the two specific electromagnetic properties of superconductors [Poo95]. Equation (2.2) shows perfect conductivity: in comparison to Ohm's law, where the current increases linearly with an electric field, in superconductors the electrons are accelerated by the electric field rather than hold their velocity against resistance [Tin96]. The second London equation (2.3) describes the exponential decay of the magnetic field on a length scale of λ. This behavior becomes obvious when equ. (2.3) is rewritten in combination with Maxwell's equation $\nabla \times \mathbf{H} = \mathbf{J} + \frac{\partial \mathbf{D}}{\partial t}$ which, in the absence of any displacement current, is Ampère's law:

$$\nabla \times \mathbf{B} = \mu_0 \mathbf{J}. \qquad (2.4)$$

2.2 The low field limit

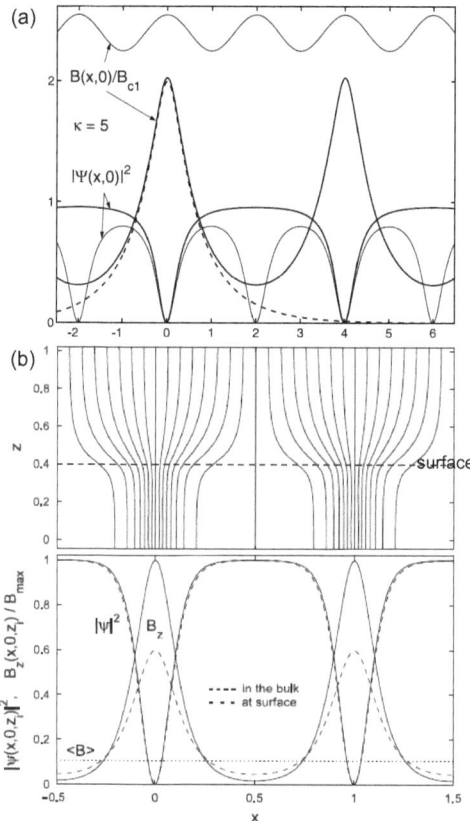

Fig. 2.3: (a) Profiles of magnetic field and order parameter for flux lines with lattice spacings of 4λ (solid lines) and 2λ (thin lines). The dashed line shows the magnetic field of an isolated vortex. (Calculations made using the GL theory by Brandt [Bra02a]). (b) Magnetic field lines above the surface (top) and profiles of order parameter and magnetic field (bottom) for a superconducting film with thickness $d = 8\lambda$ and vortex lattice spacing of 10λ calculated from GL theory [Bra09a].

2 Superconductors in magnetic fields

The resulting equation for the magnetic induction in superconductors is derived as:

$$\nabla^2 \mathbf{B} = \frac{\mathbf{B}}{\lambda^2}. \tag{2.5}$$

In the case of axial symmetry this equation can be solved by a Fourier transformation and has an exact analytical solution:

$$B(r) = \frac{\phi_0}{2\pi\lambda^2} K_0\left(\frac{r}{\lambda}\right), \tag{2.6}$$

with K_0 being a zero-order modified Bessel function [Poo95, Bra02a].

Using the two limiting forms of the function K_0 the following behavior of the magnetic induction lines in the vortex becomes clear:

$$\begin{array}{ll} B(r) \to \frac{\phi_0}{2\pi\lambda^2}\left(\frac{\pi}{2}\frac{r}{\lambda}\right)^{1/2} e^{-r/\lambda} & r \to \infty, \\ B(r) \approx \frac{\phi_0}{2\pi\lambda^2}\left(\ln\frac{\lambda}{r} + 0.12\right) & \xi \ll r \ll \lambda. \end{array} \tag{2.7}$$

These equations characterize the behavior of the magnetic field inside the superconductor. For the local studies of vortex distribution, however, the stray field distribution above the surface of the superconductor is the main object of interest. The existence of the SC/NS surface can be described by the following boundary conditions: (i) no superconducting current crosses the surface; (ii) the magnetic induction $B(r,z)$ is continuous at the film surface and satisfies $\nabla^2 B(r,z) = 0$ above the surface [Bra05]. The resulting equation for the magnetic induction lines is:

$$B_z(r,z) = \frac{\phi_0}{2\pi\lambda^2} \int_0^\infty \frac{J_0(\gamma r) exp[-\gamma(|z|-d/2)]}{k_\gamma [\coth(k_\gamma d/2) + k_\gamma/\gamma]} d\gamma \tag{2.8}$$

where $k_\gamma = (\gamma^2 + 1/\lambda^2)^{1/2}$, z is measured from the center of the film, d is the film thickness and J_0 is an integral Bessel function [Ora97, Car00]. Figure 2.3 (b) shows the magnetic field lines calculated by Brandt [Bra05] from the GL theory for a film with a thickness $d = 8\lambda$. The field decays rapidly with z so that no field can be sensed already at a distance a above the surface, where a is the vortex-vortex separation [Bra05].

Carneiro and Brandt [Car00] found out that for thick films with $d > 4\lambda$ the field distribution above the surface is similar (to first order) to the magnetic field emanated by a magnetic monopole of $2\phi_0$ located at a depth 1.27λ below the surface. Thus, equ. (2.8) is simplified in this case to:

$$\mathbf{B}(r,z) = \frac{\phi_0}{2\pi} \frac{(\mathbf{r} - \mathbf{r}_0) + (z + 1.27\lambda)\mathbf{z}}{\left((r-r_0)^2 + (z + 1.27\lambda)^2\right)^{3/2}}. \tag{2.9}$$

The quite rapid exponential decay according to equ. (2.7) of the field with a distance inside the superconductor leads only to a short-range repulsive interaction between vortices. Hence at $B \ll B_{c2}$ vortices could be considered as separated non-interacting objects. This approximation can be well applied to bulk samples and to thick films (thickness $d \gg \lambda$) where the vortex can be described as a long and thin tube. When the film thickness decreases the surface starts to play a more important role for the vortex-vortex interaction. Here vortices interact not only within the superconductor but are additionally coupled by the magnetic field outside the superconductor [Bra05] over the effective penetration depth $\Lambda = 2\lambda/d$. For very thin films ($d \ll \lambda$) the interaction between vortices occurs mostly outside the superconductor. Hence, as it was shown by Pearl [Pea64], the slow decay of the magnetic field far away from the vortex core $B(r) \propto 1/r^2$ results in the long-range magnetostatic interaction between vortices. The vortices strongly repel each other creating a regular lattice with the maximum possible distance between the next nearest neighbors. Any modification of the vortex lattice is energetically highly unfavorable.

2.3 Flux line pinning

Pinning of magnetic flux lines (Abrikosov vortices) is one of the most important requirement to a superconducting material for industrial applications. The high H_{c2} alone is not enough to ensure the capability of carrying current without losses. Assuming an ideal material without defects, when a current is applied the vortices immediately start to move under the influence of the Lorentz force in the direction perpendicular to field and current. That leads to dissipation and to a non-zero resistance. The mechanism that allows to pass current without losses is the pinning of vortices at fixed positions [Tin96]. Pinning usually happens on defects or impurities that cause a local variation of the free energy per unit length of the vortex line. The vortices prefer to be fixed (pinned) at the positions where this energetic landscape has its minima. To leave these energetically favorable positions, the vortices have to overcome a potential barrier, or, in other words, the applied force has to be higher than the pinning force and thus perform the depinning of the vortices from the defects (pinning centers). The most effective pinning occurs when the size of the pinning center has a dimension similar to ξ or λ [Tin96]. The temperature dependence of the pinning force can be written as:

$$F_p(T) = F_p(0)\left(1 - \frac{T}{T_c}\right)^n, \qquad (2.10)$$

where $F_p(0)$ varies for different materials from 10^{-12} to 4×10^{-4} N/m and n from 1.5 to 3.5 [Poo95].

2 Superconductors in magnetic fields

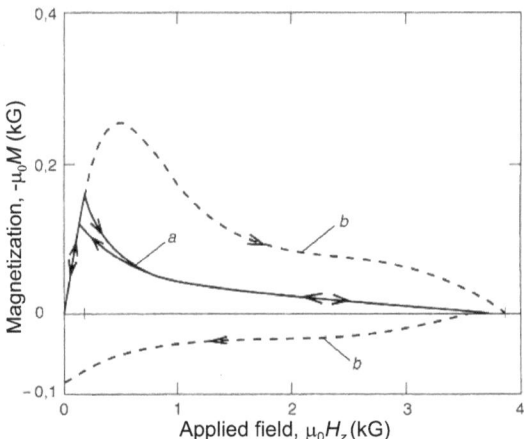

Fig. 2.4: The magnetization curve for a Nb$_{55}$Ta$_{45}$ alloy (a) without defects (b) with a high amount of defects (1 kG=0.1 T). (Adapted from [Buc04])

Natural pinning happens on local impurities, making the material properties difficult to predict. It is therefore beneficial to control the distribution of pinning centers and the pinning force artificially. This topic attracts high attention of the scientific community all over the world. However, increasing the pinning potential and, in such a way, maximizing the loss-free current density, by introducing artificial defects as well as understanding of the pinning mechanisms is up to now a challenge, both in basic and in applied research. Consequently, the local study of the pinning mechanism of individual vortices is a main topic of this thesis.

2.4 Bean's model of the critical state

When a small magnetic field is applied to a zero-field cooled superconductor the superconducting current starts to flow in a surface layer with a thickness in the order of the penetration length λ [Joo02]. This current screens the external field ensuring the Meissner state in the interior of the material. Thus, the negative magnetization increases linearly with applied field (Fig. 2.4 (a)) until the field reaches H_{c1}. For higher fields the superconductor lowers its energy by allowing the penetration of flux lines (vortices): the magnitude of the magnetization slowly decreases to zero at $H_{ext} = H_{c2}$. For ideal materials (no defects) this behavior is reversible. In reality, superconductors exhibit a hysteretic magnetization curve (Fig. 2.4 (b)). Pinning of the flux lines is the reason for a hysteretic response of the superconductor to an

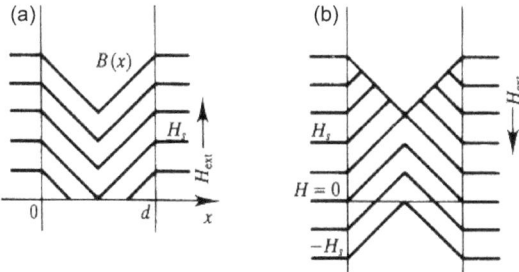

Fig. 2.5: Bean's model approximation for the internal flux-density profiles in a thick slab for (a) increasing and (b) decreasing external field H_{ext}. H_s is the maximum field that can be completely screened [Tin96].

applied field [Buc04].

Bean [Bea62, Bea64] proposed a model that could explain the experimental results and give the magnitude and the distribution of the critical current in the so-called critical state. The critical state is referred to a metastable state of the type-II superconductor, where the superconducting flux quanta, being held by the defects, cannot reach their equilibrium position [Joo02]. Bean's simple model assumes a constant magnitude of the critical current density $j_c(B) = $ constant in the volume where the magnetic flux penetrates and $j_c(B) = $ zero in the flux-free region. This assumption leads to a linear gradient of the flux density from $\mu_0 H_{ext}$ to zero inside the critical region. It turns out, that the shape of the sample is an important parameter for the quantitative application of Bean's model [Joo02]. For example, the flux-density profile in a thick slab with the slope $|\partial B_z / \partial r| = \mu_0 j_c$ is shown in Fig. 2.5 for increasing and decreasing external field. The critical current density derived from this model allows an estimation of the average pinning force density $f_p = j_c B_z$. Despite its simplicity, Bean's model is a powerful instrument and is widely and successfully used for the interpretation of experimental data.

3 Local vortex imaging techniques

> The microscope can see things the naked eye cannot,
> but the reverse is equally true.
> *Hans Selye, From dream to discovery*

For decades vortex imaging has been a topic of high interest in fundamental as well as in applied research. A huge variety of interesting phenomena is hidden in this area. Local probing is crucial for the microscopical understanding of the global properties of superconductors. For example, the dynamics of vortices in the presence of pinning centers determine the value of the critical current.

This chapter gives a short overview of different local visualization methods, their advantages and disadvantages, their limits of resolution and their fields of applications in superconductivity. Actually, almost all techniques were initially developed for room temperature applications. The difficulty of research in superconductivity is the low temperature (up to the mK range). Thus, cryosystems equipped with magnetic coils and sometimes high vacuum conditions are required.

As it is described in Chapter 2 there are two features of vortices that can be probed locally. The first one is its magnetic property: the spatial modulation of the local magnetic induction around the vortex with a maximum in the vortex core and an exponential decay over the length of the magnetic penetration depth (λ). The second one is an electronic property: the zero density of Cooper pairs inside the vortex core that scales with the GL coherence length (ξ) [Tin96]. Most local methods are based on probing the modulation of the local magnetic induction. Only scanning tunnelling microscopy/spectroscopy (STM/STS) is a technique that explores the local electronic structure, measuring the variation of the superconducting order parameter inside the vortex.

Different scattering methods, that are also widely used to study vortex lattices, such as e.g. small angle neutron scattering (SANS) [Whi08] and muon spin rotation (μSR) [Son00], as well as laser scanning microscopy [Luk08] are not presented in this short overview.

3 Local vortex imaging techniques

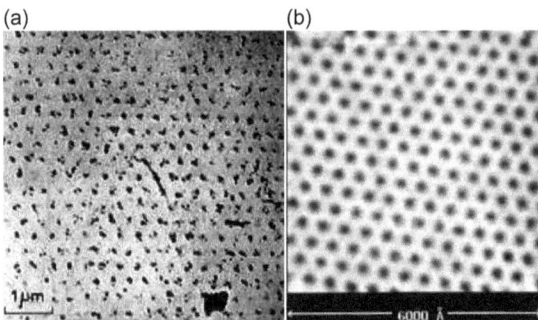

Fig. 3.1: (a) First local image of the Abrikosov lattice. Bitter decoration on Pb4-at%In rod at 1.1 K and 0.3 mT. Black points are decorated vortices [Ess67]. (b) STM image of the Abrikosov lattice in a Nb$_3$Sn single crystal at 1.8 K and 1 T [Hes89].

3.1 Bitter decoration

The existence of the Abrikosov vortex lattice was firstly experimentally confirmed by neutron scattering [Cri64] followed by nuclear magnetic resonance studies [Fit66]. These methods probe the macroscopic average over a large sample volume. The first technique that directly imaged individual flux lines and their arrangement in a "perfect" triangular lattice was Bitter decoration (Fig 3.1 (a)) [Ess67]. This highly important work experimentally confirmed the calculated magnitude of flux quanta by Abrikosov [Abr57]. In the magnetic decoration technique, developed in 1931 by Bitter to study domain walls in ferromagnets [Bit31], magnetic nanoclusters agglomerate on the surface at the positions of the magnetic field inhomogeneities. Consequently, for superconducting materials, the particles concentrate on the areas with the highest value of the stray field - at the vortex core. Afterwards, the vortex lattice can be imaged by visualizing the particles' distribution on the surface of the superconducting sample by e.g. scanning electron microscopy. This method is time consuming and the visualization of the vortex dynamics is limited by the decorating process itself. Moreover, it only provides information about the vortex positions but not detailed information about the vortex core. The sample can be decorated only once, thus for each temperature and field regime a new sample is required. Despite of a rapid development of other techniques, this method is still used for vortex imaging. The main advantage of Bitter decoration is its ability to visualize separated vortices combined with a macroscopic area of view, allowing a large statistic. Thus, it still attracts the attention of scientists studying for example the vortex distribution, their pinning and disorder in mesoscopic superconductors [Gri07], or matching of the vortex lattice with a periodic array of defects [Fas06, Bez96]. A detailed review on magnetic decoration imaging in vortex matter is given by Fasano and Menghini [Fas08].

3.2 Scanning tunnelling microscopy and spectroscopy

Scanning tunnelling microscopy and spectroscopy (STM/STS) was invented by Binnig and Rohrer in 1982 [Bin82]. The interesting fact is that they had superconducting materials in mind when they developed this technique. Until now STM is the most powerful tool to study local electronic properties of conducting materials and it has played a leading role in the experimental verification of the microscopic theory of superconductivity. This microscope has a spatial resolution down to the atomic scale, something that no other technique can do. STM is based on the quantum mechanical effect of tunneling of electrons through an isolating barrier [Wie94]. With the sample on ground, a bias voltage applied to the sharp metallic tip that scans above the surface at a distance of a few angstroms leads to a tunneling current between the sample and the tip. The current changes exponentially depending on the tip-sample distance providing an extremely high (atomic) resolution in z direction for topographical studies. However, this sensitivity is simultaneously a main drawback that restricts STM to be applicable only for very flat samples. Another limitation of STM is the conductivity of the surface. As the surface should emit and absorb electrons, a high conductivity is required. Any nonconductive precipitates as well as oxidation layers on the surface lead to the immediate crash of the tip. Thus, STM usually applies to *in-situ* cleaved single crystals with an atomically flat clean surface. The investigation of thin films is quite tricky especially for rough and, in the normal state, low conducting YBCO films (see e.g. [Loz99]).

In the spectroscopic mode (STS) the tip is fixed in one position and the tunnelling current (I) is measured as a function of bias voltage (V). Then the tip is moved to the next spatial position. The derivative $dI(V)/dV$ is proportional to the local electronic density of states near the Fermi level and, in the case of superconducting materials, reveals information about the superconducting energy gap and the GL coherence length [Wie94]. Mapping of the vortex distribution is performed by plotting the $dI(V)/dV$ for the bias voltage below the superconducting gap. For the first time the electronic structure of superconducting surface was locally probed by STS in 1984 on Nb_3Sn single crystal [Elr84] followed by the imaging of the Abrikosov lattice by Hess *et al.* [Hes89] (see Fig. 3.1 (b)). The vortex distribution can be directly correlated with topography. One example of vortex imaging on YBCO single crystals by STS is shown by Maggio-Aprile *et al.* [MA95].

STS is a very time-consuming process that demands a high stability of the microscope

3 Local vortex imaging techniques

(extreme vibration damping and stability over days and weeks of measurements). However this powerful method is an irreplaceable and extremely important method for basic research on superconductors with highest spacial and magnetic field resolution. Temperature and field dependence of the superconducting energy gap, electron bands symmetry, electronic structure of the vortex core are only few examples that can be investigated by STM/STS. The experimental highlights obtained over the last decade in HTSC by STM are nicely and systematically reviewed by Fischer *et al.* [Fis07].

3.3 Magnetic force microscopy

The idea of atomic force microscopy (AFM) came from the necessity to perform topographical studies on non-conducting samples, where STM is not applicable [Smi86]. In this technique the force which acts between the sample and the tip is measured during scanning of the tip along the surface. Thus, this method is also called scanning force microscopy (SFM). Because this interaction occurs not between two atoms (one from the tip and one from the sample), but between a large amount of atoms, atomic resolution is mostly not reached. The tip-sample interaction force is usually measured as a function of the deflection of the cantilever spring that holds the tip. In the topographical mode the force is used as a feedback signal to maintain a constant tip-sample separation. As a result, going from point to point, the tip maps the surface profile. If the tip is covered by a thin magnetic layer, the distribution of the magnetic stray field above the sample can be probed by the so-called magnetic force microscopy (MFM). This method has a high lateral resolution (down to 20 nm) and is widely used for the investigation of magnetic materials. Nowadays room temperature AFM/MFM became one of the most important devices in science and industry for studying the topography of all kinds of samples, - especially for thin films where the surface holds information about the grown mechanism and domain structures. The application of MFM at low temperatures for the vortex imaging in superconductors was firstly shown by the group of Güntherodt [Mos95]. Since then, this powerful technique was used for a variety of vortex experiments. Visualization of separated vortices, their dynamics as well as the correlation with topography were measured on all kinds of superconducting materials [Vol02, Ros02, Pi04, Lu02].

A MFM tip has his own magnetic moment [Wad92] and, thus, exhibits a magnetostatic interaction with a superconductor. According to this fact, the measured signal contains the information not only about the magnetization of the sample but depends on the precise magnetic configuration of the tip. Additionally, vortex studies appear to be not so trivial and only pinned vortices can be visualized. However, MFM gives the unique opportunity to perform a direct local manipulation of individual vortices by the MFM tip [Aus09, Str08]. The interplay between the tip and the sample has to be considered to determine the magnetic state of the sample qualitatively. To obtain further quantitative information, e.g. about the absolute stray field, the tip should be pre-characterized [Hug98, Loh00]. Nevertheless, with its high spatial resolution, MFM is an important method to study the correlation of pinned vortices with topography, as well as their depinning by the MFM tip.

3.4 Scanning Hall probe microscopy

The main advantage of scanning Hall probe microscopy (SHPM) is its ability of quantitative measurements of the perpendicular component of the magnetic field near the sample surface combined with a good spatial resolution, and non-invasiveness [Cha92]. The limitation of the spatial resolution comes from the size of the Hall sensor, that, starting from $0.85\,\mu m$ [Ora97], was rapidly improved in the mean time to $50\,nm$ [San04], approaching the resolution of MFM.

As the sensor has no magnetic or superconducting parts, the Hall microscope can be calibrated to measure the absolute value of the perpendicular component of the magnetic stray field near the sample surface with a sensitivity being between MFM and SQUID [Loz99]. Additionally, the Hall sensor does not effect the magnetic field distribution in the sample even at very small scanning distances, in contrast to MFM. Low temperature SHPM became an important tool for studying the vortex phase diagram [Ora98] and measuring the temperature dependence of the penetration length λ [Ora97]. Furthermore, because of its high scan speed (up to 8 frames/s), the real-time visualization of vortex dynamics, depinning and vortex-antivortex annihilation is possible [Ded06]. The high magnetic field resolution of SPHM allows to image individual vortices even close to T_c [Ora96]. This method is a good alternative to LT-MFM. If the distance between the scanner and the surface is controlled by tunneling, a low roughness and high conductivity of the surface are required similar to STM. In the case of AFM tracking SHPM [Cho01] this problems can be overcome. In addition, SHPM allows to correlate topography with a magnetic field distribution.

3.5 Magneto-optics

Magneto optic (MO) or Kerr microscopy is an excellent technique for local visualization of the magnetic field distribution over the sample surface in real time. Light, that falls on a magnetic medium, exhibits a rotation of its polarization plane. This effect was firstly demonstrated by Faraday in 1846. The Faraday rotation is proportional to the magnitude of the local magnetic field parallel to the light beam. Magneto-optics holds an important place in the investigations of magnetic domains, their pinning and dynamics in all kinds of magnetic materials [Hub09].

First MO images of the flux distribution in superconducting materials were obtained by Alers (1957) followed by De Sorbo (1960) with a spatial resolution of only 200 μm [Ale57, DeS60]. As superconducting materials show no Faraday or Kerr effect, a MO-active indicator film (e.g. ferrimagnetic garnet film [Han83]) is placed on top of the sample to sense the flux distribution above the surface [Dor92]. The thickness and the quality of the indicator film strongly affect the spatial resolution of the MO limiting it to about 30 μm. Despite this limitation MO has a big set of advantages. Firstly, real time movies of the flux penetration can be performed with a high spatial resolution. Secondly, any kind of samples (shape, form, thickness) can be investigated. And finally, the method is absolutely non-destructive and less sensitive to the surface quality compared to scanning techniques. Recently, a new type of high resolution MO demonstrated for the first time the ability to visualize individual vortices and their movement [Goa01]. It was shown, that the improvement of the set-up and the garnet film quality leads to a spatial resolution of about 1.3 μm. A comprehensive review of the applications of MO in superconductivity is presented by Jooss et al. [Joo02].

3.6 Lorentz microscopy

The idea to use a field-emission electron microscope [Kaw90] in the presence of a magnetic field was firstly developed and realized by the group of Tonomura (Hitachi Ltd) [Har92]. Thus, electron holography [Mat89] and Lorentz microscopy [Bon93] appeared to be new methods for the quantitative visualization of vortices and their dynamics. The electron beam, passing through a slightly tilted sample, interacts with the magnetic field of vortices. As a result, the electron beam exhibits a deflection from the preliminary direction under the influence of the Lorentz force. Therefore, each vortex is observed as a pair of black and white spots in the defocused plane of the microscope. The dynamics of vortices can be imaged in real time as well as a correlation with topography can be performed [Ton02]. This great technique provides high resolution images of vortex movements, pinning and depinning. Like in TEM investigations, the sample preparation is the big difficulty of this method. Only some 100 nm single crystals or free standing films can be studied.

3.7 Scanning SQUID microscopy

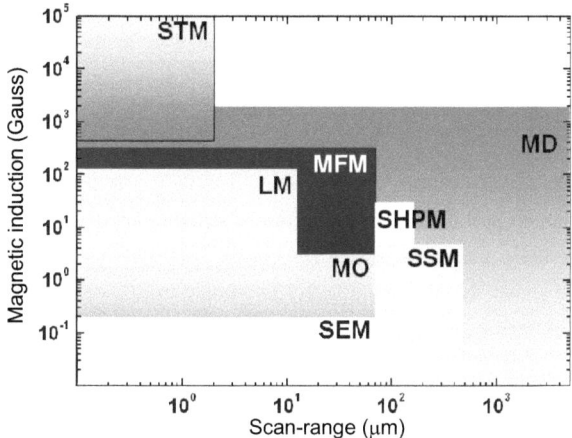

Fig. 3.2: Single-vortex resolution with different experimental techniques: range of magnetic induction versus scan-range [Fas08]. The abbreviations are: scanning tunnelling microscopy (STM), magnetic decoration (MD), Lorentz microscopy (LM), magnetic force microscopy (MFM), scanning SQUID microscopy (SSM), scanning Hall probe microscopy (SHPM), scanning electron microscopy (SEM) and magneto-optical imaging (MO). $1\,\text{G} = 0.1\,\text{mT}$.

3.7 Scanning SQUID microscopy

Scanning SQUID microscopy has a spatial resolution of about $10\,\mu\text{m}$, that is not applicable for the vizualization of separated vortices in the Abrikosov lattice. Nevertheless, the number of vortices can be extracted from the measured stray field divided by the flux quantum. This is possible because the SQUID is able to probe the absolute value of the field with a very high sensitivity – $10^{-4}\phi_0$ [Vu93]. One beautiful example of SQUID images is the visualization of a half-integer flux quantum by Kirtley *et al.* [Kir96]. Moreover, during scanning, the sensor is located at a distance of about 1 mm above the surface. Thus, the quality of the sample surface is not critical for these measurements. Important advantages of SQUID microscopy are the large scanning area and the high scanning speed. Consequently, this method is indispensable for industrial applications, e.g. for superconducting tapes and coated conductors, bringing a rapid and easy way to check their quality and homogeneity on the large scale by mapping the magnetic field distribution. As the spatial resolution is limited by the sensor size, a recently fabricated microSQUID sensor improves the resolution down to the sub-micron range [Has08].

In conclusion, Fig. 3.2 summarizes the sensitivity and scan ranges in which every technique can resolve individual vortices [Fas08]. For materials with a high λ value the physical limi-

3 Local vortex imaging techniques

tation should be taken into account. All methods that probes $H(r)$ are limited to low field values because the vortex diameter scales with λ that can be about 200 nm for HTSC. Thus, an increasing field leads to an overlapping of the vortices and herewith to problems to resolve them. As a result, STM, that probes coherence length, has the highest field resolution for such materials.

The method chosen for this thesis is low temperature magnetic force microscopy (LT MFM). It allows to perform: 1) visualization of individual vortices on relatively rough and low-conductive films; 2) comparison of the vortex distribution with the topography; and 3) investigation of vortex dynamics and vortex depinning from the defects. Thus, this technique meets all requirements.

4 Experimental methods

In der Ruhe liegt die Kraft.
German proverb

In the following chapter the experimental techniques used in this thesis are introduced. The main focus lies on the general concept and the instrumental setup of the low temperature magnetic force microscope (LT-MFM). Other methods, such as magneto-optics and room temperature atomic force microscopy are briefly mentioned.

4.1 Low temperature magnetic force microscopy

The techniques described in the previous chapter visualize the local distribution of the magnetic flux quanta in superconducting materials in different ways. All of them have several advantages and disadvantages and are preferable for one or another application. This thesis concentrates on low temperature magnetic force microscopy (LT-MFM) being the best compromise for vortex visualization in thin films and their correlation to topography. Revealing a lateral resolution of about 50 nm, LT-MFM is suitable to combine non-invasive imaging of separated vortices and their direct manipulation by the MFM tip.

General concept. The general principle of magnetic force microscopy is based on the visualization of the interaction force between a sharp magnetic tip mounted on a flexible cantilever and the stray field of the sample. The signal is recorded point by point during scanning above the sample surface. While the atomic force microscope (AFM) performs imaging of the surface topography by probing the van der Waals interaction, MFM records the distribution of the z component of the magnetic field emanating from the sample surface. Hence, to probe the magnetic landscape, the tip should possess its own magnetic moment. The standard way to produce MFM tips is to cover silicon AFM tips with a thin magnetic layer. In this thesis commercial MFMR cantilevers (Nanoworld) with a uniform Co-based coating were used.

A simplified but good approximation for the MFM tip is a magnetic monopole model [Wie94], where the tip acts as a magnetic monopole $\tilde{m}$ located at a distance δ from the physical end

4 Experimental methods

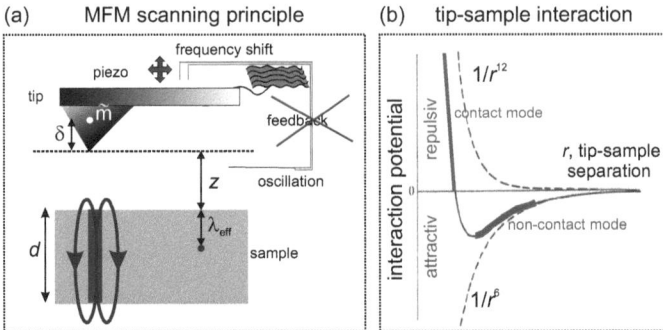

Fig. 4.1: (a) Sketch of the MFM imaging procedure. The magnetized MFM tip scans above the surface of the sample at a given distance z. The feedback loop is not active. The resonance frequency shift is detected. (b) Van der Waals force.

of the tip as it is sketched in Fig. 4.1 (a).

For the imaging of the flux distribution in superconductors low temperatures are required. The low-temperature microscope operates in high vacuum to avoid surface contamination during cooling. Moreover, vacuum gives the possibility to use the non-contact AFM operating mode during the topography scan. In comparison to the standard tapping mode where the repulsive branch of the tip-sample interaction is sensed, the non-contact mode probes the attractive Van der Waals force Fig. 4.1 (b). This ensures a maximum tip life time avoiding any contact to the surface. The constant scanning distance, which is equivalent to a constant interaction force, is controlled by the feedback loop.

To measure the magnetic contrast, after approaching the sample the feedback loop is switched off and the scan is performed at the given distance of several 10 nm from the surface (Fig. 4.1 (a)). This fact underlines the importance of having flat surfaces for the MFM studies: each precipitate with a size that exceeds the scanning height is destructive for the tip.

One additional advantage of the operation in vacuum is a strong increase of the quality factor Q of the cantilever ($Q_{vacuum}/Q_{air} \approx 500$). This improves the sensitivity and increases the signal-to-noise ratio extremely [Alb91].

Frequency modulation detection. The frequency modulation method proposed by Albrecht et al. [Alb91] is the most common technique employed in non-contact MFM nowadays. Here the MFM cantilever, that oscillates at resonance frequency, can be described as a one dimensional damped driven harmonic oscillator [Alb91]. The resonance frequency f is given

4.1 Low temperature magnetic force microscopy

by $f^2 = k_{\text{eff}}/m_{\text{eff}}$, where m_{eff} is the effective mass of the cantilever and $k_{\text{eff}} = k + \partial F_z/\partial z$ is the effective spring constant. It corresponds to the following equation of motion for the cantilever:

$$m_{eff}\ddot{z}(t) + \beta\dot{z}(t) + kz(t) + F_z(\tilde{z}) = F_{\text{ext}}, \qquad (4.1)$$

where F_z is the z component of the tip-sample interaction force and F_{ext} is the external excitation force. In the absence of the tip-sample interaction the cantilever oscillates at its own resonance frequency f_0. Via scanning above the magnetic sample, the tip undergoes an either attractive or a repulsive interaction depending on the orientation of the tip magnetization to the z component of the stray field of the sample.

In the case of small oscillations, $\Delta f \ll f_0$, the equation of motion can be linearized and has a simple analytical solution [Ros00]:

$$\Delta f = -\frac{f_0}{2k}\frac{\partial F_z}{\partial z}. \qquad (4.2)$$

Consequently, measuring the resonance frequency shift Δf caused by the interaction of the MFM tip with the stray field emanating from the sample, the tip-sample interaction force F_z can be determined. This value depends on the tip-sample separation distance. Moreover it varies from point to point during scanning above the sample, allowing to plot $x - y$ cuts of the distribution of the z component of the field gradient at a given distance above the sample.

The parameters of the MFM tip used in this thesis are presented in Table 4.1.

Tab. 4.1: Characteristic features of the MFM tip.

Tip	f_0 (kHz)	k (N/m)	Q_{air}	Δf (Hz)
MFMR, Nanoworld	75 ± 10	2.8 ± 0.5	≈ 100	0.5–5

Instrumental setup. The microscope used in this work is a commercial low temperature scanning probe microscope provided by Omicron (Cryogenic SFM). A schematic diagram of the device is shown in Fig. 4.2 (a). The upper part consists of two chambers: the microscope working chamber and the load-lock. The microscope chamber is kept in ultra-high vacuum (UHV, $< 10^{-10}$ mbar), the load-lock is used to transfer the sample to the head without breaking the vacuum. The head (Fig. 4.2 (b)) can be operated in STM and in MFM mode. The sample holder is positioned with a one-axis linear piezo-motor allowing to perform the

4 Experimental methods

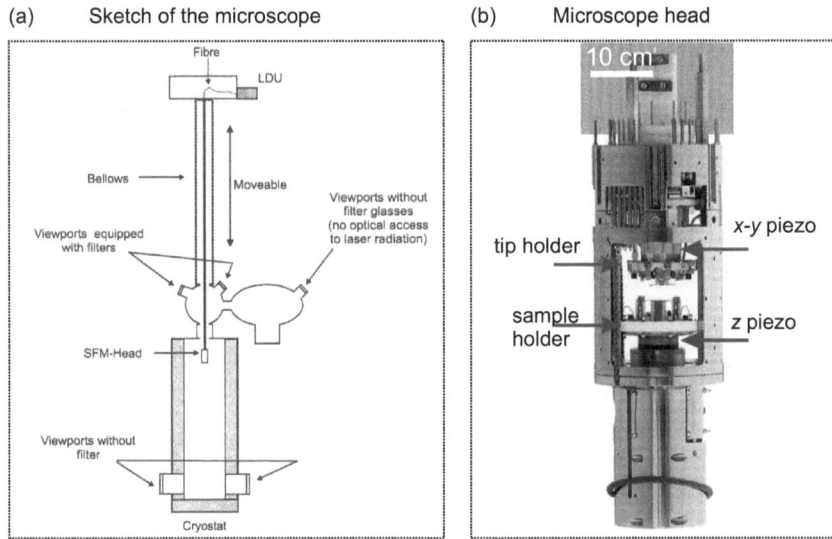

Fig. 4.2: (a) Sketch of the Cryogenic SFM (Omicron) microscope. Adapted from [Omicron user guide]. (b) Microscope head.

tip-sample approach in z direction. The cantilever coarse movement is performed by a two-axis piezo motor ($x-y$ scanning). The fine x, y, z movement during scanning is realized by the sample piezo stack. As the piezo movement depends on temperature, the maximal scanning area of 20 μm × 20 μm at room temperature reduces to about 4 μm × 4 μm at the lowest working temperature (6 K). After approaching the surface in the AFM mode, the x and the y slope of the sample is determined and the scanning distance z is set. During imaging the tip scans back and forth along the fast scan direction (usually x axis) and moves in incremental steps forward along the slow direction (y). The microscope chamber is equipped with four viewports that allow the optical access to the head for the manual approach of the tip to the sample. Because of the UHV no video camera or optical microscope can be place in the direct vicinity to the scanning head. Thus, an optical microscope is placed outside one of the viewpoints. This optical access leads to a large working distance and to problems in approaching a specified sample position. The resolution of the used optics (Leica M712(5)) is good enough to approach objects with a size of 100 μm.

The microscope cryostat is supplied with two superconducting magnets that produce magnetic fields up to 7 T vertically (out-of-plane magnet) and up to 3 T transversally (in-plane

4.1 Low temperature magnetic force microscopy

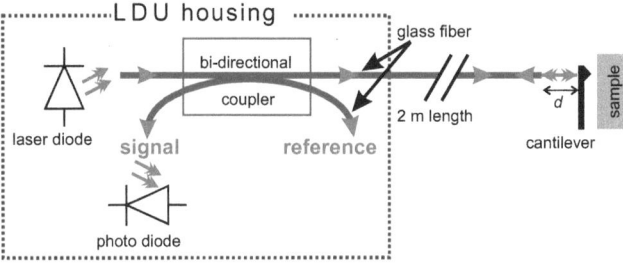

Fig. 4.3: Interferometric detection scheme. The Light/Detector Unit (LDU) (Adapted from the Omicron user guide). The measured signal intensity varies with the distance d between the cantilever and the fiber cleaved edge.

magnet). A magnetic field can be applied in 0.1 mT steps during the measurement. The LHe cryostat from Oxford provides the cryogenic temperature for the superconducting magnets (NbTi wires, $T_c = 9.5$ K and $\mu_0 H_c = 13$ T) and allows sample cooling. For temperature regulation the microscope head is in close contact with the variable temperature insert (VTI) through a long transfer rod. To reach the desired sample temperature liquid He from the cryostat is pushed into the VTI through a tiny capillary. The automatic control of the temperature is performed by an Omicron ITC 503 that controls opening and closing of a needle valve at the capillary entrance combined with the compensation of the temperature by the heater. The desired temperature of the VTI can be reached quite rapidly but, because of the UHV and the absence of the direct contact between VTI and the sample, achieving the stable desired temperature of the sample is a long term process (takes several days). The working temperature ranges from 300 K down to 6 K and the temperature stays stable in the range of 0.01 K during the measurements.

Interferometric signal detection. The force-induced deflection of the cantilever is detected by the fiber-optic interferometer [Rug89] which is sketched in Fig. 4.3. The Light/Detector Unit (LDU) is a class IIIB laser with $p = 10$ mW and $\lambda = 830$ nm. On its way to the cantilever the laser light passes through the about 2 m long glass fiber. Part of the light is reflected by the straight cleaved edge of the fiber and the rest is reflected by the back side of the cantilever and recouples again to the fiber. These two signals interfere on the way to the photo diode which is the measurement element. The resulting intensity variation is a function of the distance d (see Fig. 4.3) between the cantilever and the fiber edge: $I = I_0 + I_1 \cos(\frac{2\pi}{\lambda} \cdot 2d)$, where λ is the laser wavelength (see the Omicron user guide). The photo diode signal, after amplifying and filtering, is analyzed for its frequency with the help of a phase-locked loop (PLL). The

4 Experimental methods

PLL ensures that the cantilever always oscillates at its modified resonance frequency. In AFM mode it used as feedback signal to control the constant distance between the tip and the surface (constant height mode), so that the line which the tip follows represents the topography of the sample. In the MFM mode the shift of the resonance frequency Δf is imaged as measured signal.

4.2 Other techniques

Low temperature magneto-optics. The magneto-optical experimental setup requires a polarized light microscope, an image recording system (video camera) and a magnetic coil in order to vary the applied magnetic field. To study superconducting materials, two additional components are indispensable: (i) to set and control the temperature of the sample a cryocooling system should be used, (ii) since superconductors reveal no significant Faraday or Kerr effect, magneto-optical layers (MOL) should be placed on top of the sample [Joo02]. In this work MO imaging was performed by a modified Zeiss Jenapol polarization microscope with a 40× Long-Length objective and recorded by a low noise analog CCD camera. The cryostat Cryovac with a "cold finger" cooling principle allows to vary the sample temperature in the range from 4 K to 300 K and hold it with an accuracy better than 2 K. The out-of-plane magnetic coil with a maximum field of 18 mT and a MOL ferrite garnet film with in-plane magnetic anisotropy were used to study the field propagation into the thin superconducting stripe (Chapter 5).

Room temperature AFM. To study the topography of the films at room temperature, a Digital Instrument Dimension 3100 atomic force microscope was used in tapping mode. The thickness of the films as well as the reduction along the polished wedge (Chapter 6) was determined from AFM profile measurements along the etched step. The etching was performed with strongly diluted nitric acid. Processing of all AFM and MFM images was done with the help of WSxM free software from Nanotec [Hor07].

Transport measurements were performed in a standard four-point configuration using a 9 T Quantum Design Physical Property Measurement System (PPMS) (Chapters 5 and 7).

High resolution scanning electron microscopy (HR SEM) imaging was carried out using a Leo 1530 microscope with a Gemini electron beam unit. *Electron backscattering diffraction measurements* (EBSD) were performed with a Nordlys 2 detector using a step size of 16 nm. *Cross-sectional focused ion beam* (FIB) cuts were performed in a Zeiss 1540xB microscope (Chapter 6).

5 Vortices and defects in YBCO films

> The great tragedy of science - the slaying of
> a beautiful hypothesis by an ugly fact.
> *Thomas Henry Huxley; Biogenesis and abiogenesis*

This chapter discusses the possibility of local imaging of vortices and their correlation with artificial defects in thin $YBa_2Cu_3O_{7-\delta}$ (YBCO) films. Since the discovery of high-temperature superconductivity in cuprates by Bednorz and Müller in 1986 [Bed86], and the following sintering of YBCO with a critical temperature of 92 K [Wu87] the hope for possible industrial applications has led to various investigations on this material. Until now, YBCO coated conductors are the most promising candidates for electrical power applications or magnetic field coils. The main efforts concentrate nowadays on increasing the in-field critical current density J_c as these conductors have to sustain their own Gauss field when they carry a current or are designed to create large magnetic fields with low or zero power consumption, e.g. for medical applications. A remarkable influence of artificial defects on vortex pinning and, hence, on J_c in high fields was measured [Gut07,Hän05]. Starting from heavy ion-irradiation [Maz98], going to nanoinclusions [MD08,Eng07], columns [Kan07], substrate decoration [Mat04,Spa07], quasi-multilayers [Hän06,Bac07] and so on, this part of applied superconductivity seems to be explored the most. Nevertheless, the defects leading to an optimum pinning potential are still under investigation. Thus, the understanding of vortex pinning is indispensable for the future improvement of the performance of coated conductors. The following results were obtained in the framework of the European project HIPERCHEM[1] that deals with the modification of the critical current density by incorporation of different artificial nanodefects and at the same time with the development of cost effective and flexible deposition methods [Obr06,Kur04].

The microscopic imaging of flux lines is an alternative way to look at the vortex pinning mechanism at natural and artificial defects and to estimate the magnitude of their pinning force. First LT-MFM images of the vortex distribution in high-quality YBCO thin films were taken by Moser *et al.* in 1995 [Mos95], followed by the correlation of vortices with

---
[1] The reported results were obtained with research funding from the European Community under the Sixth Framework Programme Contract Number 516858: HIPERCHEM.

5 Vortices and defects in YBa$_2$Cu$_3$O$_{7-\delta}$ thin films

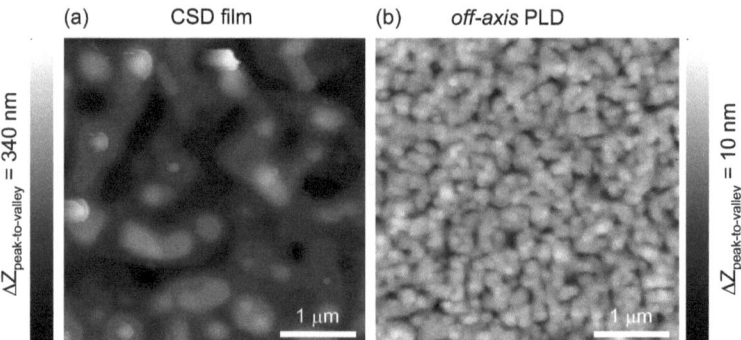

Fig. 5.1: Typical roughnesses of YBCO films measured by AFM. (a) CSD film, (b) *off-axis* PLD film. The peak-to-valley roughness of each film is shown by the scale bar. Scanning area is 4 μm × 4 μm.

natural surface defects of YBCO films [Mos98a]. The question discussed in the present work is the possibility of a direct correlation between vortices and artificial defects as well as of vortex imaging on rough as-prepared thin films. Moreover, the technical and methodological problems of this approach are depicted, proposing solutions for further research in this topic.

5.1 Surface roughness and growth process

Because of the low costs and the processing flexibility, chemical solution deposition (CSD) opens a way to future industrial applications of YBCO thin films [Cas03, Obr06]. Thus CSD films were the main investigation object of the HIPERCHEM project. Looking on these films through the eyes of a magnetic force microscope reveals an unfortunate detail: these samples appear to be absolutely unusable for local vortex imaging. As described in Chapter 4, the surface roughness for the LT-MFM studies has to be below 20 nm. Figure 5.1 (a) demonstrates the topography of a typical CSD film with a thickness of about 300 nm. The variation of the surface profile is on the same scale like the film thickness. This high porosity of the CSD films can be associated with the stress relief during the thermal treatment of the coated films [Obr06]. In this thesis a nanoscale polishing technique was developed to enable surface-sensitive studies on rough as-prepared thin films (see Chapter 6).

Compared to CSD, pulsed laser deposited (PLD) films are denser and smoother. In the standard *on-axis* geometry micron-large objects, so-called droplets, are generated by the laser plume from rough targets and agglomerate on the film surface [Kre07]. They are obstacles for the LT-MFM studies and can severely damage the MFM tip. *Off-axis* PLD deposition, where the substrate is aligned perpendicular to the target [Hol92], avoids the droplet formation and leads to films with a relatively flat surface. Thus, for local studies of vortices, *off-axis* PLD YBCO films on single crystalline (100) SrTiO$_3$ substrates have been prepared with improved smoothness by Backen and Hänisch. For that reason, different deposition conditions were investigated. In the PLD process an excimer laser with a wavelength of 248 nm and an energy density of about 2.7 J·cm^{-2} was used to ablate material which was deposited on a heated substrate. After the deposition, oxygen loading took place *in situ* during cool-down at 15 K·min^{-1} in O$_2$ atmosphere.

Oxygen pressure, target-to-substrate distance and substrate temperature have been varied. Best results were obtained using an O$_2$ pressure of 0.7 mbar, a substrate temperature of 710°C and a target-to-substrate distance of 6 cm [Hän04, Bac06]. The best achieved roughness is below 10 nm (Fig. 5.1 (b)). Increasing the substrate temperature during deposition improves the T_c and J_c values [Bac06], however, it also increases the roughness of the film. Consequently, one can not optimize J_c and surface roughness simultaneously.

The roughness of the YBCO films is based on the nature of its growth mechanism. A high density of spiral-like structures and terraces was observed at the surface of epitaxially grown YBCO films by STM [Ger91, Haw91] and AFM [Bur94] starting already from a thickness of about 10 nm. The spirals form as follows: laterally growing *ab*-terraces that coalesce incoherently create holes in between. Additionally, a vertical shift of the lattice of two growing fronts results in the appearance of 3D spirals. The unit-cell-by-unit-cell growth predominates, revealing the spirals with a step height of about 1.2 nm which correlates with the *c*-axis of the YBCO unit cell [Zhe92]. However, the not-unit-cell growth was also observed [Haa96]. The dislocation-mediated growth was shown on various substrates, for different deposition techniques and for a wide range of growth conditions [Sch94] and, thus, is an intrinsic feature of YBCO films. A detailed study of the formation of growth spirals and their evolution can be found elsewhere [Bur94, Ain94]. Such an island growth mechanism provides dislocations that act as important flux-pinning centers in YBCO films, but, at the same time, they are the main obstacles for the direct local correlation of vortices and nanoinclusions in YBCO films.

5.2 Vortex imaging on flat off-axis PLD films

The direct observation of flux lines was performed on three different smooth *off-axis* YBCO films at two different temperatures (10 K, 30 K) and at frozen magnetic fields (+2 mT, -2 mT,

5 Vortices and defects in $YBa_2Cu_3O_{7-\delta}$ thin films

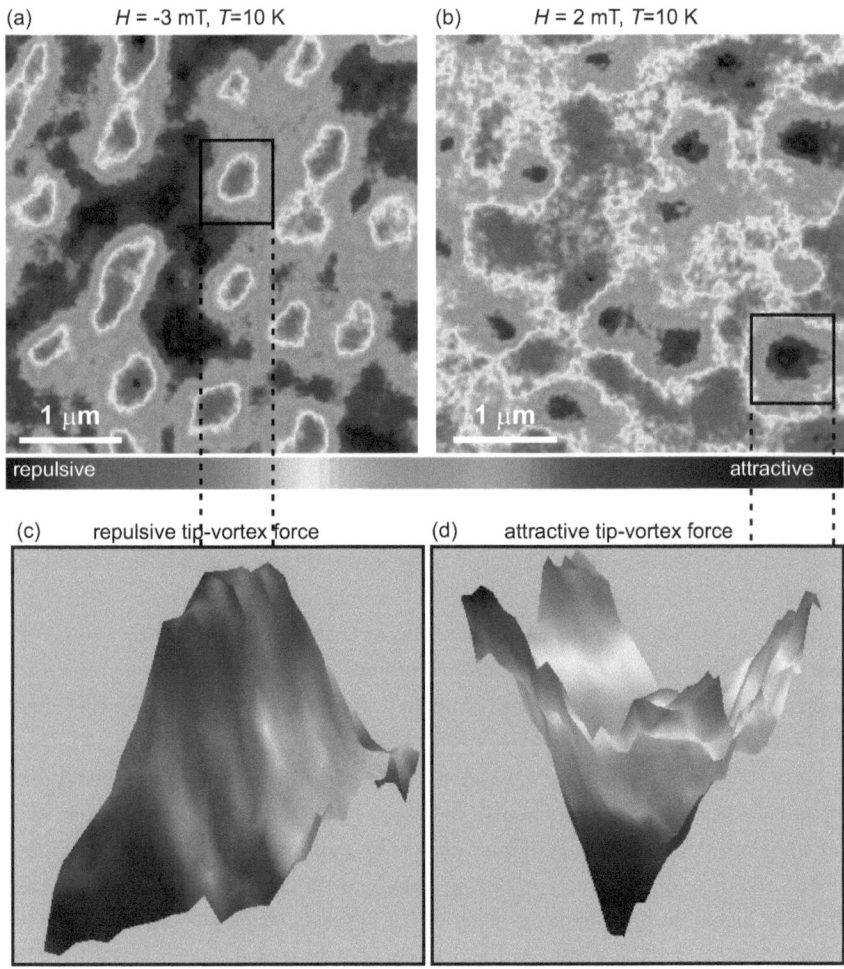

Fig. 5.2: LT-MFM images of vortices at 10 K frozen in different fields: (a) $H = -3\,\mathrm{mT}$, (b) $H = +2\,\mathrm{mT}$. The disordered vortex lattice points to a strong pinning by the natural defects in YBCO films. The scale bar represents the relative value of the measured signal: red color corresponds to a repulsive tip-vortex force and blue to an attractive one. Scanned area is $4\,\mu\mathrm{m} \times 4\,\mu\mathrm{m}$. (c) and (d) are 3D views of the individual vortices zoomed from (a) and (b) scans respectively.

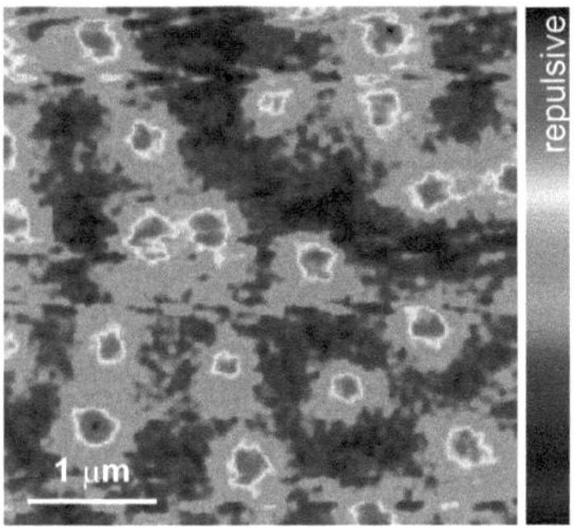

5.2 Vortex imaging on flat off-axis PLD films

H = -2 mT, T = 30 K

Fig. 5.3: The LT-MFM image of vortices at 30 K and -2 mT.

-3 mT). The reproducibility of the experiment was confirmed. As it was described in Chapter 2, the magnetic field penetrates the superconductor in the form of separated flux quanta $\phi_0 = 2.07 \cdot 10^{-15} \mathrm{m}^2 \cdot \mathrm{T} = 2.07 \mu\mathrm{m}^2 \cdot \mathrm{mT}$, ordered, in the ideal case, in the hexagonal Abrikosov lattice. The number of vortices is proportional to the applied field and can be easily counted to be 24 for 3 mT and 16 for 2 mT at a scanning area of 4 μm × 4 μm.

As explained in Chapter 4, the signal measured by MFM is a frequency shift of the oscillating scanning tip caused by the stray field above the sample (equ. 4.2). This signal includes a non-trivial interplay of tip and sample magnetization. Consequently, in this chapter the measured MFM contrast should be interpreted only qualitatively. This means that the vortices are visualized as red spots (repulsive interaction) or blue spots (attractive interaction) depending on the direction of the frozen field relative to the MFM tip magnetization. A more quantitative analysis of MFM scans is performed in Chapters 7 and 8 on LTSC thin films.

Measuring a homogeneous 200 nm thick film (Fig. 5.1 (b)) with an average peak-to-valley roughness of 10 nm and a frozen-in field of -3 mT reveals vortices as separated round objects with positive frequency shift (red spots) (Fig. 5.2 (a)). It was verified, that if the direction of the frozen field changes, the magnetic field of the flux lines exerts an attractive force on the MFM tip and vortices are imaged as blue spots (Fig. 5.2 (b)). The number of imaged vortices correlates with the expected values for the hexagonal Abrikosov lattice, and the vortex arrangement is stable over several weeks of measurements. This confirms that vortices are not manipulated by the stray field of the tip.

The average vortex-vortex distance at $H = -3\,\mathrm{mT}$ (Fig. 5.2 (a)) is estimated to be about 850 nm which correlates with the expected value for the hexagonal vortex lattice $a = \sqrt{\frac{2}{\sqrt{3}} \frac{\phi_0}{B}} \approx$

5 Vortices and defects in YBa$_2$Cu$_3$O$_{7-\delta}$ thin films

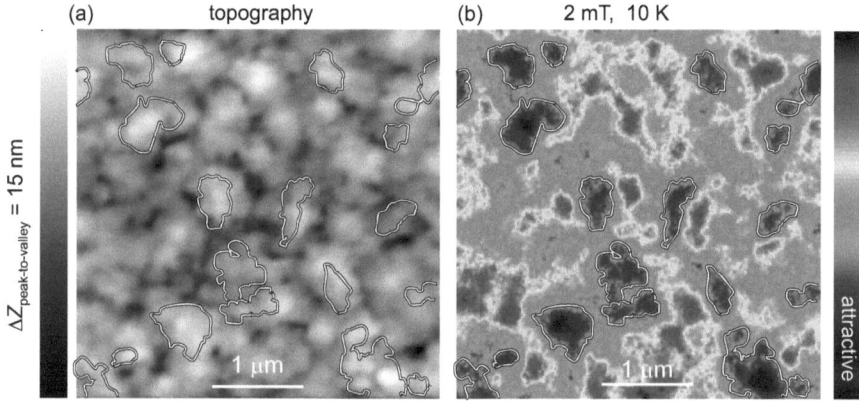

Fig. 5.4: (a) AFM image of the *off-axis* PLD film with slightly increased roughness. (b) LT-MFM image on the same position, $H = 2\,\text{mT}$, $T = 10\,\text{K}$. White contours depict flux lines imaged as blue objects. The scanned area is 4 μm × 4 μm.

$1.075\sqrt{\frac{\phi_0}{B}}$, ($B = \mu_0 H$ with H being the applied field). At low temperatures ($T = 10\,\text{K}$) and low fields, the average displacement of the vortices from an ideal lattice position (ca. 130 nm) is caused by the pinning of vortices by grain boundaries (intergrain pinning) [Fer03,Mos98a], leading to the glass-like pattern of the flux lines, the so-called topologically ordered Bragg glass [Gia97]. The image at $H = -2\,\text{mT}$ and $T = 30\,\text{K}$ (Fig. 5.3) reveals a smaller displacement of the vortices from the ideal lattice positions compared to the lower temperature and the average vortex-vortex distance of 1200 nm. Also at this slightly elevated temperature the intergrain pinning is still strong enough to prevent a movement of vortices by the MFM tip.

If the surface roughness increases (Fig. 5.4 (a)), vortices are not separated anymore but are imaged as conglomerated flux lines with a fuzzy shape (Fig. 5.4 (b)). Here, the positions of the flux lines, depicted by the white contours, are compared with the topography. No exact statement about the preferable location of flux lines at the surface corrugations can be given. Volume pinning seems to play an important role for the rough films. Due to the fact that no literature data are available about the local correlation of vortices with artificial nanodefects in YBCO films, this question is discussed will be Section 5.4.

5.3 Local and global measurements of the critical current

There are different ways to measure the critical current density in superconducting materials. Global methods, such as voltage-current $U(I)$ curves and magnetisation measurements,

5.3 Local and global measurements of the critical current

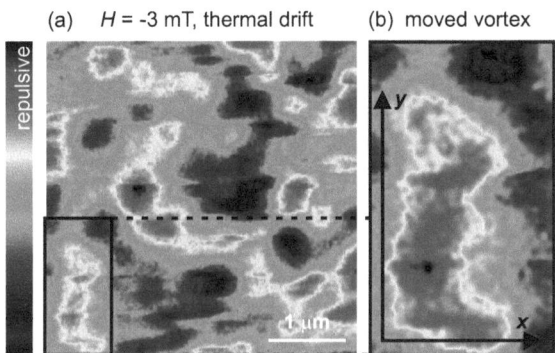

Fig. 5.5: (a) Thermally activated drift of vortices, $T = 10\,\text{K}$, $H = -3\,\text{mT}$. The scanned area is $4\,\mu\text{m} \times 4\,\mu\text{m}$. (b) Zoom-in of the vortex elongated in the y direction during the scanning process.

provide average values of the pinning force, while local techniques give access to the pinning and depinning at individual defects, evaluating their pinning potential. A local study of the flux penetration in the presence of defects was performed by magneto-optics [Joo02] on a μm-scale, but no local studies of the current-induced movement of individual vortices exist in the literature.

For transport measurements *off-axis* PLD YBCO films have to be structured. Bridges of $50\,\mu\text{m}$ and $100\,\mu\text{m}$ width and $3\,\text{mm}$ in length have been patterned by optical lithography and ion-beam etching and provided with low ohmic contacts by gold evaporation followed by a heat treatment ($300°\text{C}$, $30\,\text{min}$, $400\,\text{mbar}$ O_2). The application of very small currents in the microscope during measurements leads to a vortex movement via scanning (Fig. 5.5 (a)) that is not caused by the applied current ($I \ll I_c$), but is due to the temperature drift coming from the Joule heating of the silver contacts on the positions where the wires touch the sample. The MFM tip scans back and forth along the x axis and moves in incremental steps forward along the y axis. Figure 5.5 (b) shows a zoom-in of the vortex elongated in the y direction during the scanning process. A recent publication by Auslaender et al. presents a beautiful study of the controlled movement of vortices in an YBCO single crystal with an MFM tip [Aus09]. This example demonstrates the high potential of the MFM for vortex matter investigations as well as for the direct single-vortex manipulation. The thermally activated depinning of vortices under the influence of the magnetic moment of the MFM tip is discussed in Chapter 7 for the NbN thin film.

From the $U(I)$ curves in zero field and at different temperatures, the critical current is found to be $I_c = 0.3$ A at 10 K (Fig. 5.6 (a)). Increasing the temperature leads to a decrease of the pinning force (equ. 2.10) and, thus, to lower I_c values. This I_c, as already stated above, represents an average value of the pinning force and includes the effect of a huge amount of defects across the whole transport bridge.

For local studies current pulses (10 pulses with a profile: 0.1 s *on* + 0.2 s *off*) with an amplitude from 0.1 A to 1 A were applied to the sample[2], and the MFM scans were performed after the temperature of the sample was stabilized [Sha07]. As it is shown in Fig. 5.6 (b) and (c) MFM images in the middle of the bridge reveal no movement of the flux lines up to 1 A, which is largely above the critical current estimated from the global transport measurement. The reason for this disagreement was clarified by magneto-optical studies and is presented below.

Magneto-optical imaging was performed as described in Section 4.2. The sample was cooled down in zero field to 10 K and an out-of-plane field was applied. Figure 5.6 (d) shows the MO images along the 3 mm bridge. It is clearly seen that the bridge contains several defects on the left and on the right side. Only the middle part is shown to be relatively homogeneous. Consequently, during transport measurements, the voltage drops dramatically over those defects that have higher resistivity than the homogeneous film. Hence, the local critical current can be much higher than the average (global) one, measured during the transport experiments. This fact explains the contradiction between global and local transport measurements. MO investigations on different YBCO films lead to the same results.

Additionally, low temperature MO is a powerful instrument widely used to study the field penetration, the dynamics of flux lines and their interaction with defects in superconducting materials (see e.g. review by Joos et al. [Joo02]). Based on the simple Bean model (Chapter 2), for example, the critical current density for the thin film stripe can be estimated as:

$$J_c = \frac{\pi H_p}{d} \frac{1}{\ln(\frac{2W}{d})}, \qquad (5.1)$$

with W being the stripe width, d the stripe thickness and H_p the external applied field that leads to the penetration of magnetic flux to the center of the stripe (penetration field) [Bra96]. For our stripe (Fig. 5.6 (f)) with $W = 100\,\mu$m and $d = 300$ nm the critical current density is thus: $J_c = (\pi \cdot 17.2 \text{ mT})/(\mu_0 \cdot 1.95\ \mu\text{m}) = 2.2 \cdot 10^6 \frac{A}{cm^2}$. The resulting value is usually much higher than the globally measured one [Joo02]. The critical current densities J_c measured in three different ways are presented in Table 5.1.

To conclude, the local imaging of vortices on such inhomogeneous films give no represen-

---

[2] the highest current that can be technically applied to the microscope head is 1 A

5.3 Local and global measurements of the critical current

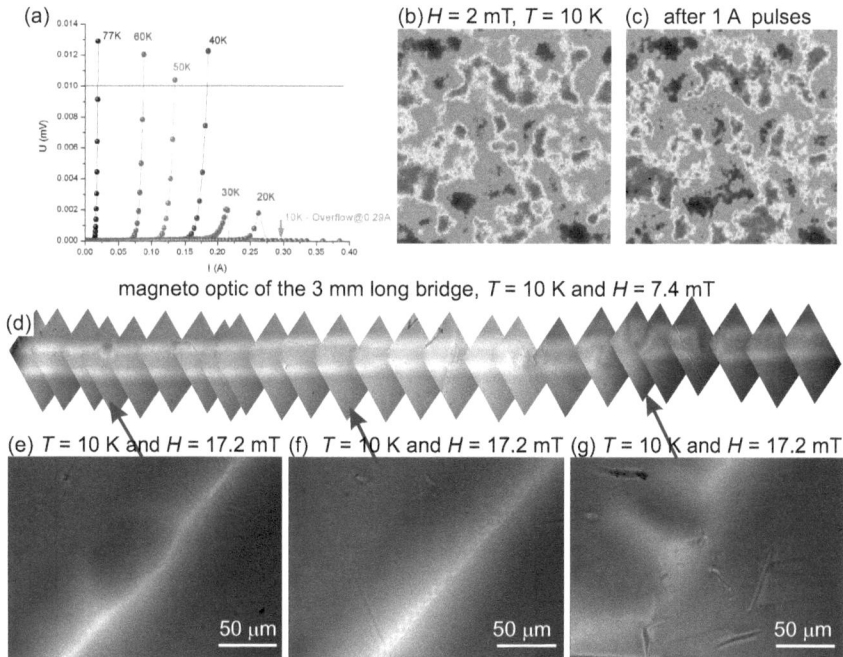

Fig. 5.6: (a) $U(I)$ curves measured on the structured YBCO bridge (3 mm long, 100 µm wide and 300 nm thick) at zero field and at different temperatures. The critical current is estimated to be 0.3 A at 10 K. (b) LT-MFM images on the bridge at +2 mT and 10 K. The flux lines are blue objects with a fuzzy shape. (c) LT-MFM scan on the same position after current pulses with an amplitude of 1 A have been passed through. Despite the fact that the current is largely above I_c, the flux distribution did not change. (d) Magneto-optical images of the whole bridge at 10 K and 7.4 mT. (f) The same sample measured at 17.2 mT in the middle of the bridge. (e) and (g) measured at 17.2 mT on the left and right defects. The field penetrates through the whole bridge.

5 Vortices and defects in YBa$_2$Cu$_3$O$_{7-\delta}$ thin films

Tab. 5.1: Critical current density (J_c) evaluated for a thin YBCO stripe by different methods

Method	$U(I)$	MO	MFM
J_c, 10^6 A/cm^2	1.0	2.2	>3.3

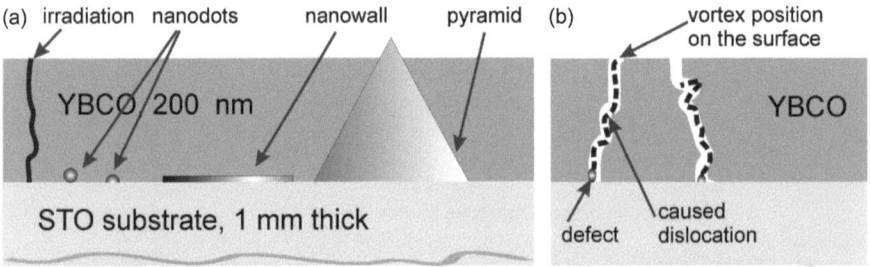

Fig. 5.7: (a) Sketched representation of all defects mentioned in this section. (b) Sketch shows the possible shift between the vortex visible on the surface and the nanodefect that pin it. Vortex is shown by the white line.

tative result for the whole film. To image a current induced flux movement by LT-MFM, a better sample homogeneity is needed, which was not reached in the YBCO films on hand.

5.4 Artificial defects

Enhanced pinning of vortices on artificial defects with a size of about the coherence length ξ significantly improves the critical current density in high fields [Tin96, Hau04]. The following section asks the question, if it is possible to correlate vortices with different artificial defects on the local scale. As shown in Section 5.2, the vortex imaging was successfully performed on different YBCO films. Thus, the main problem is to recognize artificial nanodefects on the surface topography of the YBCO film by AFM imaging and then to compare it to the MFM data from the same position on the sample. Two ways to incorporate pinning centers into YBCO are discussed: substrate decoration and irradiation. The defects are sketched in Fig. 5.7 (a). For each kind of defects presented below an increase of J_c was observed by $U(I)$ measurements, pointing to the high pinning potential of these defects.

5.4 Artificial defects

Substrate decoration

One possible way to create nanodefects in superconducting films is to use a substrate decoration method [Mat04, Spa07, Ayt08].

Y_2O_3 *nanoparticles from the gas phase.* Spherical Y_2O_3 particles with a mean diameter of 10 nm were deposited by Sparing on $SrTiO_3$ (STO) substrates by DC magnetron sputtering from an Yttrium target in an Ar gas atmosphere at an elevated pressure of 1.5 mbar. During this process the particles nucleate homogeneously and grow from the supersaturated vapor prior to the deposition onto the STO substrate [Spa07]. It was shown that the particles did not migrate on the substrate during a heat treatment similar to the YBCO deposition conditions [Spa07]. Unfortunately for local vortex studies, no signature of the nanoparticles was observed on the surface after deposition of a very smooth (5 nm peak-to-valley) YBCO layer. This effect of "covering" of the underlying defects happens due to the 3D spiral like growth mechanism of YBCO films. Thus, it was impossible to correlate vortices with this kind of defects. A similar problem is expected for all kinds of artificial nanodefects incorporated in YBCO films.

CeO_2 *nanodots and nanowalls.* Other investigated systems were CeO_2 and $Ce_{0.9}Gd_{0.1}O_2$ nanodots and nanowalls grown on STO and LAO ($LaAlO_3$) substrates by CSD based on self-organization processes by ICMAB Barcelona [Gib07]. A specified position on the template was marked and the particle distribution was imaged by AFM. Then, a very smooth thin YBCO film was deposited and the marked position was retraced again in the AFM. The comparison of these two topographical images shows that with a smooth (5 nm peak-to-valley) and 80 nm thick film one cannot see any signature of nanodots with a mean height of 10 nm on the surface of the YBCO film. Thus no correlation between nanodefects of this size and the magnetic flux distribution could be established. Further decreasing of the YBCO thickness is not useful because for film thicknesses below λ ($d \ll \lambda$) the interaction of vortices is determined by the effective penetration length $\Lambda = 2\lambda^2/d$, which is larger than λ [Bra05] leading to an earlier overlapping of the vortices and a decreasing contrast. Thus, the film thickness is also a limiting factor for the local vortex imaging on thin films.

A 100 nm *off-axis* YBCO thin film was deposited on a template with nanowalls. It was found that nanowalls disturb the growth process of the film, creating holes around them. It is assumed that the full magnetic flux penetrates into the sample through these holes and thus such samples are not acceptable for vortex imaging.

$La_{0.7}Sr_{0.3}MnO_3$ *pyramids.* Off-axis PLD YBCO films were deposited on two templates decorated with $La_{0.7}Sr_{0.3}MnO_3$ pyramids by ICMAB Barcelona [Mor09]. The maximal height of the pyramids on the first template was about 250 nm. Thus, 200 nm YBCO has been

5 Vortices and defects in $YBa_2Cu_3O_{7-\delta}$ thin films

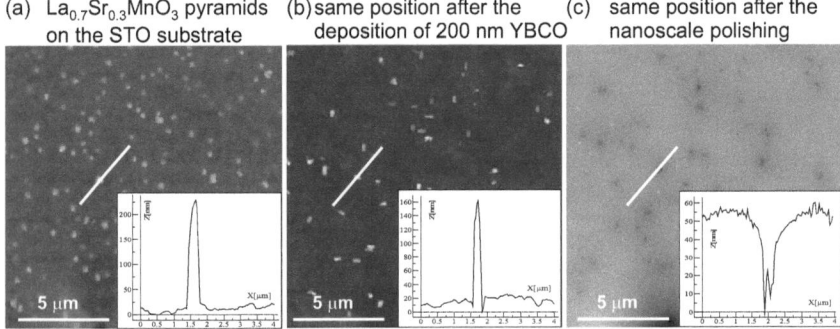

Fig. 5.8: Room temperature AFM images (15 µm × 15 µm) of (a) $La_{0.7}Sr_{0.3}MnO_3$ pyramids on the STO substrate. (b) The same position after the deposition of a flat 200 nm thick YBCO film. (c) The same position after nanoscale polishing. The insets show the line profiles.

deposited in order to correlate the flux distribution with the artificial defects. The migration stability of the pyramids was checked by AFM after heating the templates under YBCO deposition conditions. Figure 5.8 (a) and (b) show the same position on the STO substrate before and after the deposition of YBCO. AFM measurements after deposition reveal that YBCO was not only distributed between the defects, but also grows on the top of the pyramids. Such agglomerates result in a rough surface morphology, which is destructive for the MFM tip. Therefore, nanoscale polishing, described in the next chapter was applied to obtain a flat surface. Unfortunately, the initial positions of the pyramids could not be recognized in the AFM image after polishing (Fig. 5.8 (c)). Scanning electron microscopic (SEM) images and point Energy Dispersive X-ray Analysis (EDX) show that the objects that look like small holes on the SEM scan reveal a higher concentration of La than the YBCO matrix. Also focused ion beam (FIB) cross-cuts on these positions show a pyramid-like object underneath the YBCO film [The07]. Nevertheless, there was no possibility to perform MFM and FIB investigations at the same position of the film (the reason is described in the next section). The correlation of vortices even with these quite large defects was again not successful.

The second template consists of smaller $La_{0.7}Sr_{0.3}MnO_3$ pyramids (maximal height of about 100 nm). A 80 nm thick YBCO film was deposited on this template. Although the film was deposited using the same deposition conditions like the previous one, the YBCO appeared to be not a flat film, but a discontinuous layer with pores and holes throughout the whole thickness. Therefore, this film was not suitable for vortex imaging.

Irradiation

Irradiation provides columnar defects which penetrate through the whole film thickness causing a high pinning potential [Hol93, Kra93, Civ97].

Heavy ion irradiation. Heavy ion irradiation was performed by Fuchs on a preliminary polished *off-axis* PLD film to create amorphous regions of about 20 nm diameter acting as artificial columnar defects [Kra93]. 1080 MeV Pb ions with a fluence of $5 \cdot 10^7$ ions/cm^2 were used in this experiment [Fuc06]. The whole sample was irradiated causing a random but homogeneous distribution of the defects. Thus, about eight defects on the scanning area of 4 μm × 4 μm are expected. Unfortunately, the created defects were not visible at the surface by AFM, making the correlation of artificial defects with vortices in LT-MFM impossible. One possible way to see these defects on the surface is High Resolution Electron Backscatter Diffraction (HR EBSD) that allows to recognize the amorphous traces on the film surface.

FIB holes. By means of FIB ordered arrays of holes with a diameter of 20 nm and a period of 1 μm have been patterned in 200 nm thick YBCO films by Thersleff [The07]. The holes are easily recognizable on the surface by AFM scans. Vortex imaging on periodically ordered antidots usually shows well known commensurate pinning effects [Mos96, Shi08]. Unfortunately, this nanostructuring process has no real industrial application for coated conductors because of its high time and cost consume.

5.5 Is there any possibility to correlate vortices with defects?

The discussion above demonstrates a large number of attempts to locally study the pinning mechanism in YBCO films. During the last two decades a variety of different defects was empirically introduced into YBCO films that significantly improve J_c in high fields. Nevertheless, a full understanding of the pinning mechanism on the microscopic scale in this well studied material is still missing. The reason, as discussed in Section 5.1, lies in the material itself. It is not trivial to measure vortices on the surface of YBCO films because of their high roughness. But the main feature that makes a correlation almost impossible is to find traces of the nanodefects on the surface of the films, either within the LT scanning force microscope or with other external characterization techniques (e.g. HR EBSD), owing to the small scaled nature of the defects. Despite this problem could not be solved with the present state of technology, the methodology developed is applicable for future work since surface sensitive techniques and nanomanipulation are further improving.

5 Vortices and defects in $YBa_2Cu_3O_{7-\delta}$ thin films

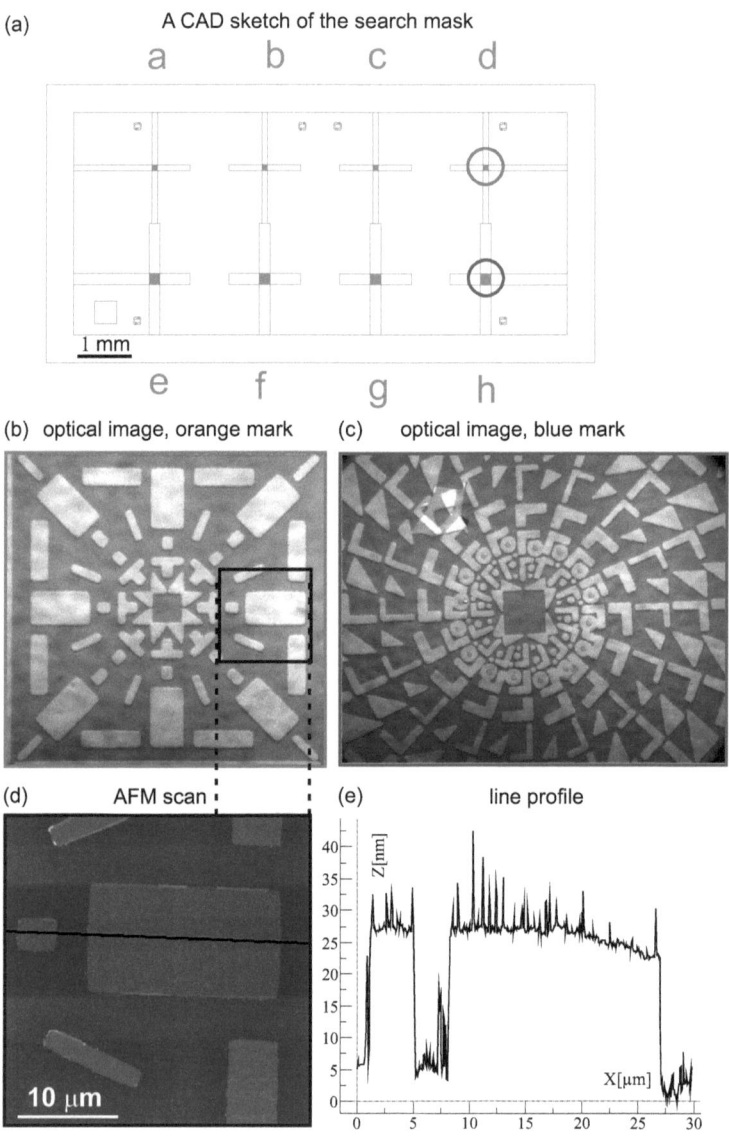

Fig. 5.9: (a) CAD sketch of the search mask for the 5 mm×10 mm sample. The crosses are visible at the surface of the film during manual approach. (b) and (c) are optical microscope images at the positions marked by orange and blue color, respectively. (d) 30 μm × 30 μm AFM scan at the position marked with a black square in picture (b). (e) Line profile of the AFM scan.

5.5 Is there any possibility to correlate vortices with defects?

Two important developments were achieved in the course of this work, which facilitate the local imaging of flux lines and will be helpful for future attempts to correlate vortices with pinning defects: 1) a unique nano-scale wedge polishing of thin films, introduced in the next chapter; 2) a specially designed lithographical search mask. The necessity of such a mask came from the fact that the low temperature force microscope usually works in ultra-high vacuum to avoid impurity adsorbation on the cold surface and to increase sensitivity. Thus, the direct video access to the sample is limited, and the scanning can only be realized at randomly selected surface positions (usually somewhere in the center). The scanning probe microscope used in this work (see Chapter 4), is equipped with an optical microscope that allows selecting a scanning area on the sample with a lateral precision of $100\,\mu m$. Thus, the search of a pre-determined position of a smaller size is impossible. Consequently, no direct correlation could be established between the measurements performed with LT-MFM and other local techniques.

A mask, which overcomes this limitation, was realized in two steps by optical and e-beam lithography by Mönch [Mön07]. A 20 nm thin gold layer is deposited onto the sample by magnetron sputtering or by PLD, followed by a lift-off process (Fig. 5.9). The optic-lithographically prepared crosses with a width of $100\,\mu m$ and $200\,\mu m$ depict the position of interest and can be approached manually. The small structures were prepared by e-beam lithography. They are organized in such a way that by scanning one AFM image with an area of $20\,\mu m \times 20\,\mu m$ one can deduce exactly how close the center of the cross is and how many steps are required to reach the position of interest ($10\,\mu m \times 10\,\mu m$ or $20\,\mu m \times 20\,\mu m$ respectively). This mask can be used for any kind of thin films and for different high vacuum microscopes with limitations in the optical access. Moreover, it opens a way to compare the scans taken on the same sample position by different surface-sensitive local studies, for example MFM and EBSD.

Like the name "surface-sensitive technique" suggests, only the surface of the sample is probed by these methods, while the interior of the film stays hidden. Due to the fact that vortices can be bent by defects [Bla94], as it is sketched on Fig. 5.7 (b), the correlation of the flux visible on the surface of the film with the defect that acts as the respective pinning center becomes even more complicated. For example, by substrate decoration, pinning of vortices occurs not at the nanoparticles but at columnar dislocations provoked by these particles during film growth [Mat04]. A way to get insight into the distribution of such dislocations is e.g. HR TEM [Zur07, MD08, Hor08].

In conclusion, despite the fact that no correlation of vortices and artificial defects could be shown, the methodology developed here can be directly applied to any other superconducting material with lower natural pinning and better homogeneity and will be helpful for future studies on coated conductors. Following the described approach one can compare the impact of different defects on vortex pinning and determine their pinning potential locally.

6 Nanoscale wedge polishing of thin films

The requirement for high surface quality of thin films for the LT-MFM study leads to the development of the technique which is presented in this chapter. The quest for the possibility to perform vortex imaging on YBCO films with a high as-prepared roughness as well as to try to avoid droplets and precipitates on the surface of the deposited films (see chapter 5) provokes looking for suitable polishing methods. As a result, an innovative nanoscale polishing has been developed that is not limited to YBCO thin films but can be used for various kinds of materials.

First of all, it is shown that upon polishing the roughness of a sample can be reduced to a few nanometers. Additionally, not only the planarity of the film can be improved but also the polishing angle can be varied up to 10 angle-seconds. In this way it opens the unique possibility to look into the interior of submicron thin films with different surface-sensitive scanning techniques, such as scanning electron microscopy (SEM), high resolution electron backscatter diffraction (HR EBSD) and any kind of scanning probe microscopy (SPM) such as atomic force microscopy (AFM), magnetic force microscopy (MFM) or scanning tunneling microscopy/spectroscopy (STM/STS).

This nanoscale wedge polishing of thin films was applied, as an example, for studying the geometry and distribution of artificial nanodefects within YBCO films. As artificial defects and inclusions have an enormous influence on the properties of superconducting materials (see chapter 5), this information is crucial for a further improvement of the superconducting properties of coated conductors. Other examples of the application of the polishing procedure are discussed in Section 6.4.

At present, transmission electron microscopy (TEM) and focused ion beam (FIB) cross-section cuts are state-of-the art techniques to look at the cross-sectional structure of a film. Both of them provide a highly resolved thickness-dependent information, but are limited by a small scan area. Another way to investigate the volume distribution of inhomogeneities is a 3D EBSD-FIB technique, where the inside of a sample is explored by removing thin surface slices of material between individual measurements [Kon06,Fel05]. All these methods are very precise and powerful, but the sample preparation and measurement procedures are complex

6 Nanoscale wedge polishing of thin films

and time- and cost-consuming. Therefore, it is of interest to have a simple and cheap method that gives a quick and reliable depth-sensitive profiling of thin films.

Wedge polishing is well known from TEM sample preparation, but up to now not widely explored for other microscopic investigations. An independent study of Solovyov et al. [Sol07] recently showed an important and very useful application of low-angle polishing for thick YBCO tapes. They used metallographic polishing which gives a rapid assessment of the texture quality of micrometer thick layers using optical microscopy and XRD analysis. The nanoscale mechanical polishing of a sample wedge, developed and applied here, is also applicable to thick films but the method developed during this work is very innovative in the nm-range, where the surface quality is decisive for subsequent microscopic investigations [Sha08]. Moreover, a very large area (from hundreds of microns to several millimeters) can be polished and prepared for a surface-sensitive analysis of inhomogeneities inside the film. The method allows a lateral scanning through the whole thickness profile of a film and, thus, reveals the information from the interior of the film to the surface.

6.1 Experimental details

Dealing with film thicknesses in the sub-micron range, the most significant aspect of the polishing procedure is the ability to mount the sample parallel to the polishing towel, thereby allowing a homogeneous reduction of the film thickness. The second step is a tilting of the film with an accuracy of a few angle-seconds to produce the necessary film thickness reduction in wedge form. The experimental setup is based on the Logitech PP5D Precision Polishing Jig, mainly used in semiconductor industry, and known for its fine polishing quality. The head was modified and upgraded with micrometer screws to adjust a precise angle setting. The samples were glued onto a glass plate using quartz wax (Ocon-200, Logitech) and were vacuum mounted on the head. The polishing machine PM5 from Logitech allows a variation of the rotating speed from 1 to 70 rotations per minute with a loading weight from 50 to 2800 g. For soft and very thin films, the parameters were selected as: 400 g, 20 rotations per minute for 5 minutes. As the YBCO films are sensitive to water, a polishing solution (OP-AN neutral alumina suspension, Struers, pH 7-7.5, 0.02 μm) dissolved in red lubricant (0.5 g in 50 ml), was prepared similar to the procedure of Wada et al. [Wad03]. The polishing velvet towel (chemomet F-towel Buehler or chemocloth Logitech) was wetted in advance to avoid the adhesion of the head. After polishing, the film was cleaned for 5 min by using the same polishing procedure but with pure red lubricant as a cleaning solution and a new polishing towel.

6.2 Plane polishing of YBCO films

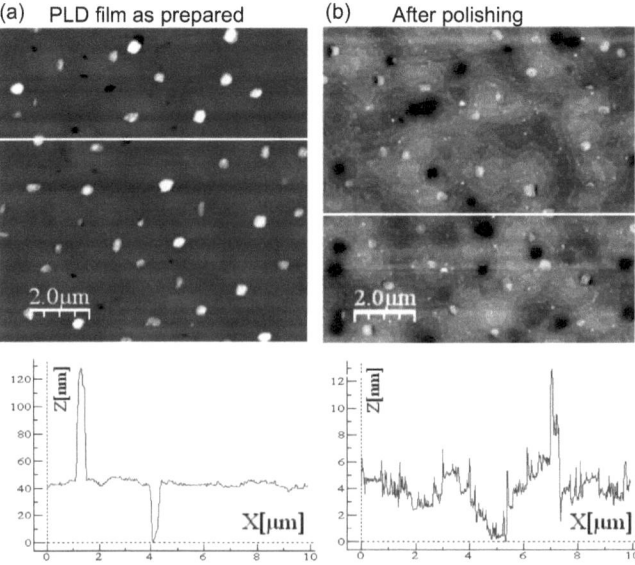

Fig. 6.1: 10 μm × 10 μm AFM image of a PLD YBCO film (a) as-prepared and (b) after planar polishing.

200 nm thin YBCO films prepared by *off-axis* pulsed laser deposition (see Chapter 5) and chemical solution deposition (CSD) on $SrTiO_3(100)$ single-crystalline substrates were used for the polishing experiments. The CSD film was produced in the trifluoroacetic acid (TFA) process by Engel. A TFA-YBCO precursor solution was prepared using stoichiometric YBCO powder dissolved in propionic acid and trifluoroacetic acid in a molar ratio of 8:1 at a temperature of T = 120°C. After the evaporation of the solvent at 140°C, the TFA precursor was dissolved in acetone and propionic acid. The Hf concentration was adjusted to 1 at.% Hf regarding to stoichiometric $YBa_2Cu_3O_7$ powder by dissolving Hf(IV)-2.4-pentanedionate in the solution. After dip coating the $SrTiO_3(001)$ substrate and drying the film, the heat treatment was completed in a temperature regime described by Falter *et al.* [Fal02]. Further details of the sample preparation can be found elsewhere [Bac07, Eng07].

6 Nanoscale wedge polishing of thin films

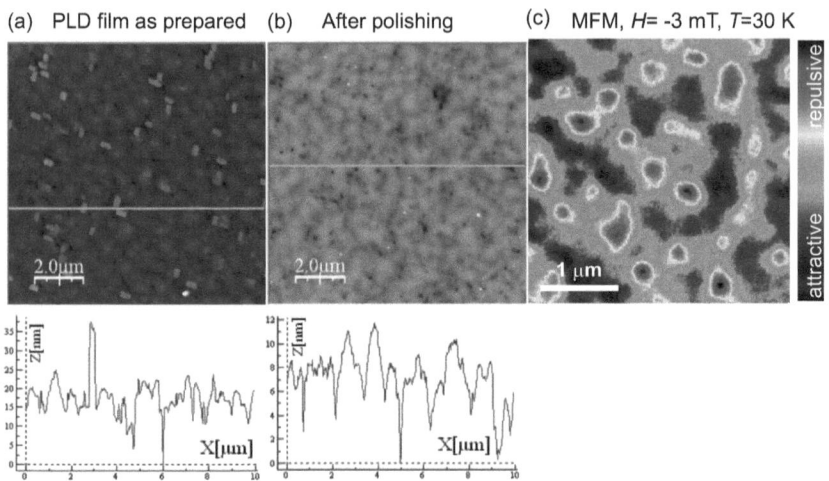

Fig. 6.2: 10 μm × 10 μm AFM image of a PLD YBCO film (a) as prepared – 35 nm high precipitates are seen, (b) after planar polishing – roughness about 10 nm and (c) 4 μm × 4 μm LT-MFM image after polishing: -0.3 mT, 30 K, in-field cooling.

6.2 Plane polishing of YBCO films

Figure 6.1 shows AFM images of an *off-axis* PLD YBCO film as-prepared (a) and after planar mechanical polishing (b). The as-grown film contains numerous precipitates, several of them more than 100 nm in height, which present detrimental obstacles for sensitive scanning techniques. After polishing, the roughness over the whole film surface has been improved to less than 5 nm (peak-to-valley) and even growth spirals of YBCO can be clearly seen. Subsequent transport measurements reveal that the polishing procedure does not lead to severe degradation of the global superconducting properties.

For example, $T_c(50\%)$ and ΔT_c of this *off-axis* PLD film measured inductively before and after polishing were 88.6 K 1.1 K and 88.45 K 1.2 K, respectively. Figure 6.2 is a demonstration of vortex imaging after reducing the roughness to below 10 nm (peak-to-valley) via planar polishing. The 200 nm *off-axis* PLD film contains precipitates of about 30 nm diameter which can be seen on the surface via AFM (Fig. 6.2 (a)). This drastically restricts or even prohibits a SPM investigation of such a sample. After polishing, the detrimental precipitates were removed and the roughness of this film decreased to 10 nm (Fig. 6.2 (b)). The film was cooled to 30 K in a field of -3 mT and the shift in resonance frequency upon scanning with an MFM tip at a distance of 40 nm above the sample surface was recorded (see Chapter 4). Vortices

6.3 Wedge polishing of nano-engineered YBCO films

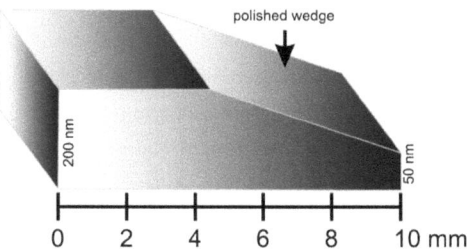

Fig. 6.3: Sketch of a polished wedge with a reduction of film thickness from 200 to 50 nm on a 5 mm length.

are clearly visible as red spots (Fig. 6.2 (c)). The number of frozen vortices observed in the 4 μm × 4 μm scan range corresponds to the number expected theoretically for a flux line lattice forming in the applied field (see Chapter 2). Thus, the developed polishing procedure opens up a unique way to perform local studies not only on selected single crystals or perfectly flat thin film samples but on largely applicable material such as coated conductors. The successful LT-MFM image of frozen magnetic flux (Fig. 6.2 (c)) after polishing additionally confirms that polishing has no serious influence on the physical properties of the film.

6.3 Wedge polishing of nano-engineered YBCO films

The above described experiments made use of the roughness reduction via precision polishing in a planar geometry. In contrast to this, mounting the sample under a defined angle additionally results in a monotonic thickness reduction of the film along the length of the sample and thus creates a smooth surface at any depth within the film. Figure 6.3 shows a sketch of a polished wedge with an angle of about 20°, which corresponds to a reduction of the film thickness from 200 to 50 nm over a length of about 5 mm. Wedge polishing was performed on a 200 nm thick Hf-doped YBCO film grown by Engel on a single-crystalline $SrTiO_3$ substrate by CSD described above [Eng07]. Since nanosized $BaHfO_3$ particles drastically improve the pinning force density and hence increase J_c and the irreversibility field in CSD YBCO [Eng07], a detailed investigation of artificial inhomogeneities is crucial to correlate the pinning properties with the microstructure.

To demonstrate the power of wedge polishing, Fig. 6.4 shows high resolution SEM images made by Engel at different positions along the wedge. Figure 6.4 (a) and (b) have been taken at a distance of 3 mm and 4 mm from the thick edge of the sample and show the non-polished

6 Nanoscale wedge polishing of thin films

part of the film. Figure 6.4 (c)–(f) illustrate the roughness changes along the polished wedge in steps of 1 mm when measured from the thick edge toward the thin one. The obvious film densification and porosity reduction with decreasing thickness along the wedge is thought to be a result of the specific way of growth during chemical solution deposition. At a thickness below 100 nm the YBCO film is dense and reveals almost no pores. Due to the chemical growth method, a 200 nm thick film becomes always rough and porous (Fig. 6.4 (a) and (b)). Consequently, with HR EBSD, it is not possible to identify the crystal structure or grain orientation on an as-prepared film at the positions of deeper holes, so that these areas appear as black, non-indexed regions within the fully indexed surface map (Fig. 6.5 (a)).

In contrast, Fig. 6.5 (b) shows an HR EBSD map measured on the thinnest part of the wedge (corresponding to the SEM image (Fig. 6.4 (f))). The surface is optimally flat and contains no non-indexed parts. As it was estimated from a cross-sectional SEM image by Engel et al. [Eng07], the nanoparticles have a lateral extension between 10 and 50 nm. A similar size distribution for PLD-prepared $BaHfO_3$ particles was shown by Backen et al. [Bac07].

The EBSD measurements were performed using a grid size of $16 \times 16 \, nm^2$. Most of the nanoparticles have been identified as nonsuperconducting $BaHfO_3$ and are smaller than 20 nm. Thus, they appear as a single pixel in the EBSD map (Fig. 6.5, blue points). The different colors in Fig. 6.5 (b) belong to different mismatch angles between $BaHfO_3$ particles and the YBCO matrix. The small non-superconductive inclusions are distributed homogeneously and show a clear epitaxial relationship to the YBCO matrix. Larger particles have random orientation and are agglomerated in large clusters surrounded by pure YBCO (white color). The density of small epitaxially grown $BaHfO_3$ particles can be estimated to be about $100 \, \mu m^{-2}$ which is equal to 2 vol%. The amount of Hf introduced during the preparation corresponds to 5 vol%. It is assumed that the remaining part of Hf is accumulated in large clusters with random orientation.

Pole figures were calculated from the measured EBSD maps. Fig. 6.6 (a) shows the (100) pole figure of the pure YBCO film. From the (111) $BaHfO_3$ pole figure (Fig. 6.6 (b)) with four different poles at a tilt of 55° one can clearly see the above-mentioned epitaxial relationship of 20–30 nm small perovskite $BaHfO_3$ particles in the YBCO matrix. The non-cubic textured component seen as speckled background intensity in the pole figure (Fig 6.6 (b)) is a result of large, randomly distributed clusters of $BaHfO_3$ precipitates.

Based on the EBSD images shown in Fig. 6.5 (a) and (b), the homogeneous distribution of the non-superconductive precipitates throughout the whole film thickness is proven.

The reported wedge polishing method provides not only an indepth analysis of thin films but also helps to correlate different surface-sensitive and position-resolved measurements.

6.3 Wedge polishing of nano-engineered YBCO films

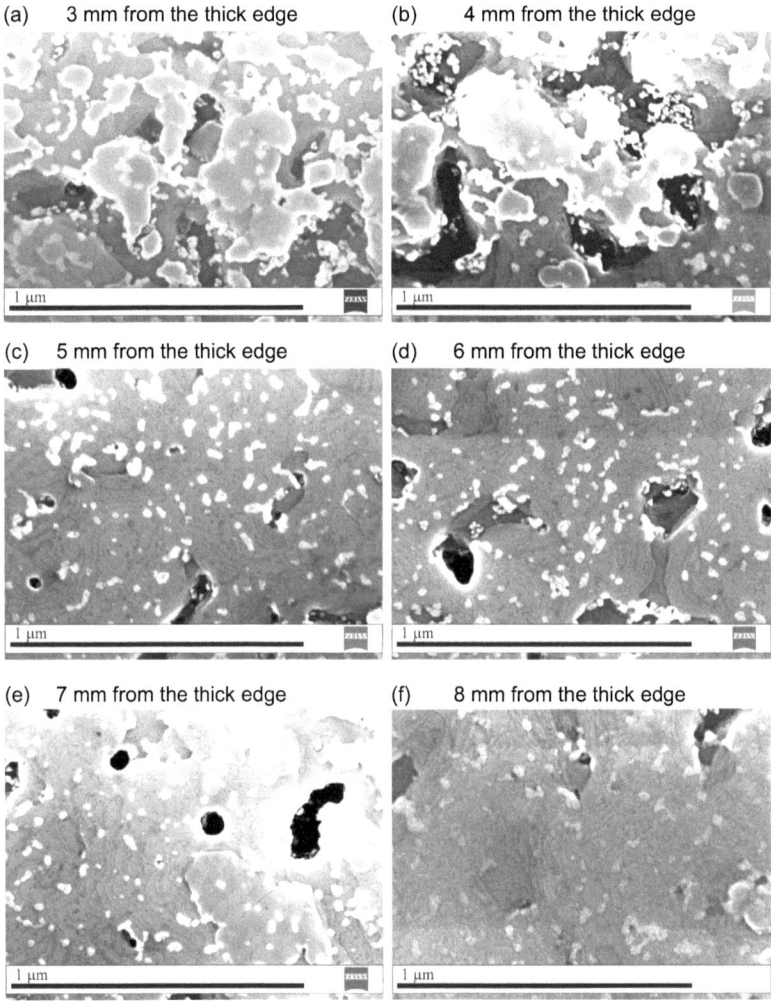

Fig. 6.4: High resolution SEM images along the polished wedge measured at a distance of (a) 3 mm, (b) 4 mm, (c) 5 mm, (d) 6 mm, (e) 7 mm and (f) 8 mm from the thick edge of the sample.

6 Nanoscale wedge polishing of thin films

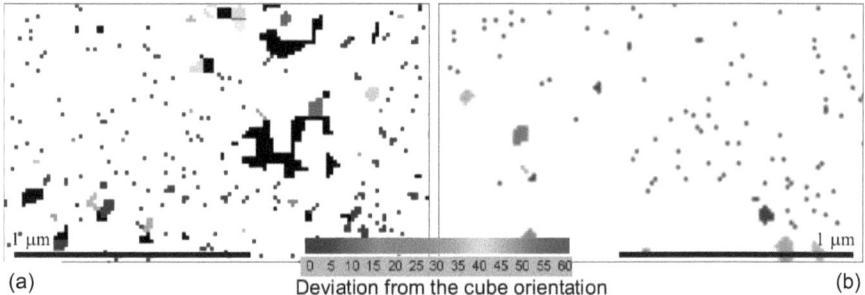

Fig. 6.5: HR EBSD (a) on the as-prepared film and (b) on the thinnest part of the wedge. Black represents non-indexed regions. White color corresponds to the YBCO matrix, blue (small points) corresponds to the epitaxially grown 20–30 nm small $BaHfO_3$ precipitates, all other colors (larger points) belong to the different mismatch angle between $BaHfO_3$ particles and the YBCO matrix.

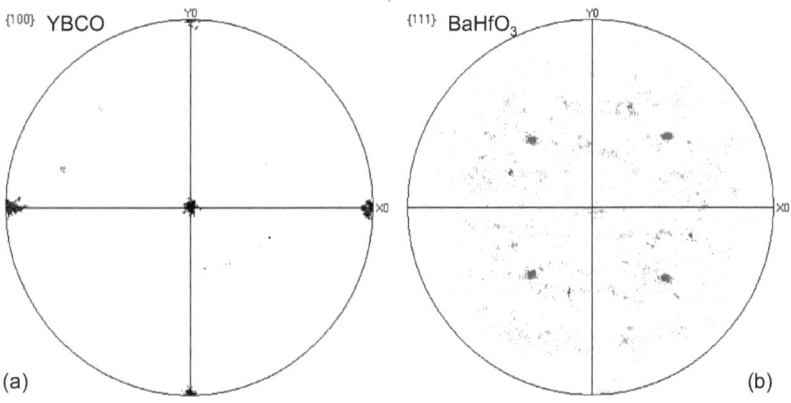

Fig. 6.6: (a) (100) pole figure of the pure YBCO film. (b) (111) pole figure of $BaHfO_3$. One can clearly see the epitaxial relationship of the perovskite particles to the YBCO matrix. The non-cubic textured components in pole figure (b) come from randomly distributed clusters of precipitates. Pole figures were calculated from measured EBSD maps.

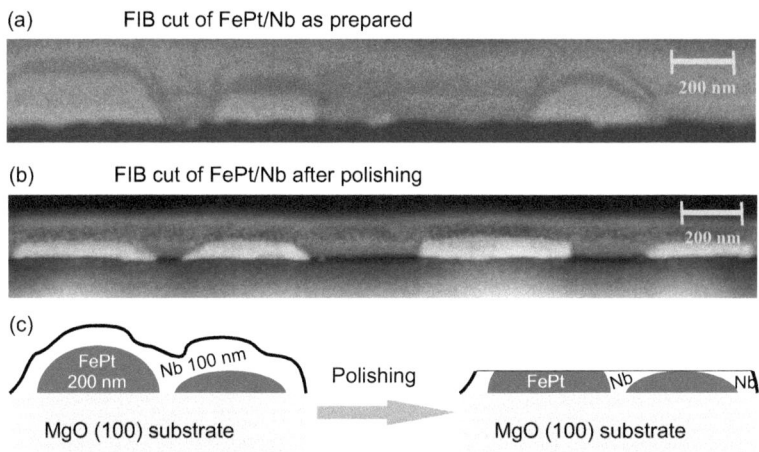

Fig. 6.7: Cross-sectional FIB cut of (a) an as-grown and (b) a polished FePt/Nb double-layer. (c) Sketch of the achieved topology.

6.4 Other applications

The nanoscale polishing is an important tool allowing to improve the surface quality of different thin films and has been applied e.g. to NbN, Bi, LaSrCoO films [Mei09]. The precipitates and droplets, that appear during the deposition process on the surface of these films were successfully removed. The polishing parameters (weight, speed, time) were selected depending on the hardness of the film and on the goal of polishing. Regeneration and cleaning of the films after degradation of the surface were also performed by polishing. Additionally wedge polishing can be used to look in the interior of heterogeneous nano-engineered thin films as well as multilayer films, which play an important role in many areas of research and industry.

An interesting demonstration of the polishing technique was the creation of a unique topology in hard magnetic/superconducting heterostructures. FePt/Nb thin films were deposited via PLD as described by Haindl et al. [Hai08]. The epitaxial growth of the FePt $L1_0$ phase on MgO(100) substrates results in separated islands with a sub micron size [Wei04]. Conventional geometries of FM/SC layered structures are either magnetic dots located on top of the superconducting film [Lan03, Vil08] or a superconducting film deposited on top of the FM structure [Hof08] (see also Chapter 8). Figure 6.7 shows cross-sectional FIB cuts of the film before and after polishing. The as-prepared structure consists of FePt islands covered by the 100 nm Nb film as shown in Fig. 6.7 (a). After polishing (Fig. 6.7 (b)) the Nb layer on

6 Nanoscale wedge polishing of thin films

the top of the islands is removed leading to a geometrical confinement of the superconducting material to areas in between the ferromagnetic islands, as it is sketched in Fig. 6.7 (c). This new topology is an interesting object of investigation [Hai09] and can be used for other heterostructures, too.

In conclusion, this chapter demonstrates that the developed nanoscale polishing is an alternative approach to investigate the internal structure of thin films and can provide a continuous insight from the film surface down to the substrate. The main challenge of this work is to polish a wedge of a very thin film (200 nm) with an angle of a few angle-seconds.

7 Pinning investigation in NbN thin film

> What we observe is not nature itself,
> but nature exposed to our mode of questioning.
> *Werner Heisenberg*

While the previous two chapters have been devoted to studies on structural aspects of superconducting thin films, the following chapter covers fundamental research on the pinning mechanism of separated magnetic flux lines in conventional superconductors. The interpretation of pinning mechanisms, the search for artificial defects with high pinning potential and commensurable pinning effects by ordered arrays of defects initiated a large variety of studies, which are today an important topic in basic research as well as in application-based engineering. The challenging task for local imaging techniques is, therefore, to determine a correlation between the positions of superconducting vortices with microstructural defects and, in addition, to investigate depinning of the flux lines from different defects. The fact that a MFM scanning tip possesses its own magnetic moment (Chapter 4) is simultaneously a challenge for a quantitative interpretation of the images, but also a unique opportunity to combine the non-invasive vortex imaging with a local manipulation of the separated flux quanta.

7.1 Tip-vortex interaction: monopole model

As discussed in Chapter 4 the magnetic moment of the MFM tip can be well approximated to first order by a magnetic monopole with a magnetic charge $\tilde{m}$ located at a distance δ from the sharp end of the tip pyramid (Fig. 4.1 (a)). Moreover, the field distribution from a single flux quantum measured just above the surface of the superconductor is similar to the magnetic field emanated by a magnetic monopole of $2\phi_0$ located at a depth of 1.27λ below the surface [Car00] (Chapter 2). Due to this fact, the tip-vortex interaction force $\mathbf{F}(r,z)$ can be estimated using a monopole-monopole model:

$$\mathbf{F}(r,z) = \tilde{m}\mathbf{B}(r,z), \tag{7.1}$$

7 Pinning investigation in NbN thin films

where $\mathbf{B}(r, z)$ is the magnetic induction of the vortex at the distance z above the surface (see equation 2.9). This model works optimally for z and $\delta \gg \lambda$ [Str08]. Taking into account, that the signal measured by MFM (Δf) is proportional to the z-derivative of the force $\mathbf{F}(r, z)$ that acts between the tip and the sample (see equation 4.2), the expression for the measured shift of the resonance frequency Δf of the cantilever can be written:

$$\Delta f = -\frac{f_0}{2k} \frac{\tilde{m}\phi_0}{2\pi} \frac{(r-r_0)^2 - 2(z+1.27\lambda+\delta)^2}{((r-r_0)^2 + (z+1.27\lambda+\delta)^2)^{5/2}}, \qquad (7.2)$$

where r_0 is the position of the vortex core and r is a radial distance from the center of the vortex. As the magnetic induction of the vortex is maximal at the center and decays exponentially with r (see Fig. 2.3 (a) and equation 2.7), the strongest interaction in z direction between tip and vortex occurs when the tip passes the center of the vortex ($r = r_0$). Thus, the maximum vertical value of the force is:

$$\max(F_z)_{r=r_0} = \frac{\tilde{m}\phi_0}{2\pi} \frac{1}{(z+1.27\lambda+\delta)^2} \qquad (7.3)$$

The ratio between the lateral and the vertical force that acts during scanning varies from 0.3 for a tip with a less sharp pyramid [Wad92] to $2/3\sqrt{3} \approx 0.38$ for the monopole-monopole model [Str08]. This results in a maximum lateral component of the tip-vortex interaction force given by:

$$\max(F_{\text{lat}}) \approx 0.38 \cdot \max(F_z) \qquad (7.4)$$

Non-invasive imaging of vortices by MFM is possible only if the vortices are pinned. The tip-vortex interaction force can be accurately tuned during scanning by varying the tip-sample separation z [Str08, Aus09]. If this force exceeds the pinning force of an individual vortex at a natural or artificial defect the vortex can be dragged from its position. Keeping the tip-sample distance constant, a change in temperature can lead to a local depinning of the individual flux lines during scanning, since the pinning force is a function of temperature equ. (2.10).

This part of the thesis is focused on the estimation of pinning forces in conventional superconducting films. In order to compare the local and the global pinning forces qualitatively and quantitatively, extended measurements on NbN thin films were performed. These thin films with T_c=16 K and a thickness of about 100 nm were fabricated by Engelmann and Haindl on single crystal MgO(100) substrates by pulsed laser deposition using a Nb (99.95%) target in nitrogen gas atmosphere ($p_{N_2} = 5 \times 10^{-2}$ mbar) [Hai09]. As on-axis PLD leads to the formation of droplets and precipitates at the surface [Kre07], the developed polishing

7.2 Local depinning of individual flux lines

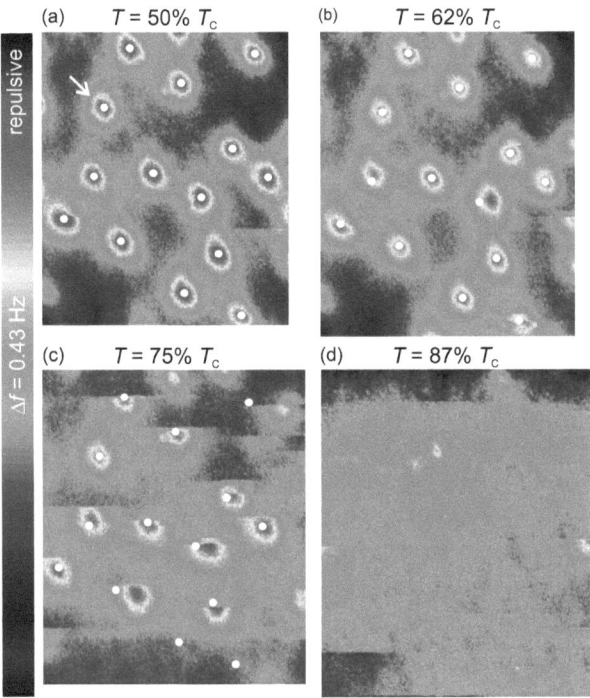

Fig. 7.1: Vortex imaging on NbN film, field cooled in $-3\,\mathrm{mT}$ at (a) $50\%T_c = 8\,\mathrm{K}$, (b) $62\%T_c = 10\,\mathrm{K}$ and (c) $75\%T_c = 12\,\mathrm{K}$ (d) $87\%T_c = 14\,\mathrm{K}$. Scanning area: $4\,\mu\mathrm{m} \times 4\,\mu\mathrm{m}$, scanning distance $z = 30\,\mathrm{nm}$.

technique [Sha08] (Chapter 6) was successfully applied to remove these obstacles providing a peak-to-valley roughness below 5 nm.

7.2 Local depinning of individual flux lines

The temperature dependence of the vortex distribution was imaged after field cooling in a vertical applied field $\mu_0 H_z = -3\,\mathrm{mT}$ (Fig. 7.1) beginning at 8 K and then increasing the temperature in steps of 2 K. As the MFM tip and a vortex exhibit a repulsive interaction, the vortices are imaged as red objects. The tip-sample distance was kept stable at $z = 30\,\mathrm{nm}$ providing a constant tip-vortex interaction force (7.3). The white points depict the positions of the vortices at $50\%T_c$ allowing to trace the rearrangement of flux lines with

57

7 Pinning investigation in NbN thin films

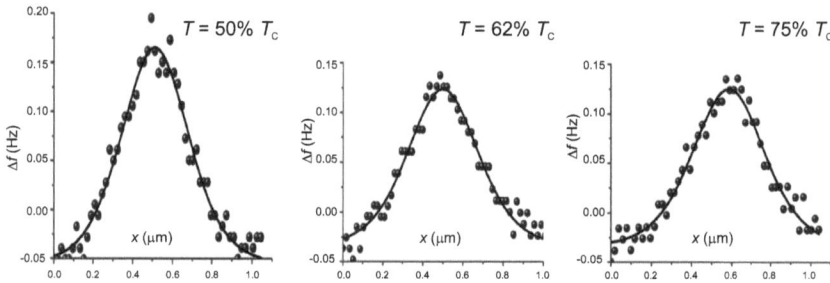

Fig. 7.2: Temperature dependence of the vortex profile. Blue points are the measured signal, the solid line is a fit with the monopole-monopole model.

increasing temperature. The vortices in NbN have sharper profiles in comparison to YBCO films (Fig. 5.2) and are better separated from each other. This is due to a better homogeneity of the NbN films and a lower surface corrugation.

At temperatures below $50\%T_c = 8\,\mathrm{K}$ the vortices are organized in a slightly disordered Abrikosov lattice (vortex glass) due to pinning by natural defects (Fig. 7.1 (a)). At this temperature a non-invasive imaging of vortices is possible, indicating that the pinning force is larger than the dragging force of the MFM tip.

The decreasing contrast of the vortex profile at higher temperatures (Fig. 7.1 (b)) agrees with the temperature dependence of the penetration length λ that is responsible for the field decay inside the vortex (Chapter 2). Most vortices are still in their original positions, thus, at a slightly increased temperature of $62\,\%T_c$ the tip-vortex interaction force is still lower than the mean pinning force. Only 2 out of 14 visualized vortices are dragged from the frozen position (white spots) to the nearest pinning sites with a higher pinning potential. This indicates the existence of a slightly modulated pinning landscape in the NbN film and, hence, a spatial variation of the pinning force.

At $75\,\%T_c$ a movement of the vortices by the MFM tip can be observed (Fig. 7.1 (c)). Consequently, at this temperature the pinning force for almost all vortices is equal to the maximum lateral force the MFM tip exerts onto the vortex (equ. 7.4). As a result, such vortices are imaged as half "cut" object. Only one vortex (depicted by the arrow) remains stable indicating the locally enhanced pinning force at this position. Three vortices are fully pushed away already when the tip crosses their field lines. At higher temperatures (Fig. 7.1 (d)) vortices can no longer be imaged. Here the vortices are continuously dragged by the tip during scanning.

For a quantitative analysis, the profiles of all imaged vortices were fitted using the monopole-monopole model described by equation (7.2). As an example, the fit of a selected vortex, depicted by the white arrow (Fig. 7.1 (a)), is shown in Fig. 7.2 at different temperatures. The temperature-independent value of the tip-vortex force (equ. 7.4) was calculated to:

$$\max(F_{\text{lat}}) = 0.74 \pm 0.04 \, \text{pN}. \tag{7.5}$$

The temperature dependence of the monopole-monopole distance $z + 1.27\lambda + \delta$ is shown in Table 7.1.

Tab. 7.1: Temperature dependence of the tip-sample distance in the monopole-monopole model

Temperature (K)	8	10	12	14
$z + 1.27\lambda + \delta$ (μm)	0.31	0.40	0.42	-

An increase of this value with temperature is related to the temperature dependence of the penetration length λ (Section 2.1). As δ is an unknown parameter of the tip, an exact value of λ can not be derived with this set of measurements. More statistics are required to extract the temperature dependence of the penetration length, $\lambda(T)$, with an accuracy of 10-20% from the MFM images [Ros01]. To get an absolut value of λ the MFM tip should be pre-characterized preliminaryly which would allow to determine its intrinsic magnetic parameters δ and $\tilde{m}$ quantitatively [Hug98, Loh00]. The so far known penetration depth for NbN films is $\lambda(0) \approx 200$ nm [Kub84, Wan96]. This value is in good agreement with the values obtained here (Table 7.1).

7.3 Global estimation of the pinning force

While the MFM provides access to the pinning force of separated vortices, the global characterization methods, such as transport or magnetization measurements, explore the collective behavior of the flux line lattice. They evaluate the mean pinning force within the whole sample volume, considering also the elastic interaction between individual flux lines as well as the collective pinning [Tin96]. For small magnetic fields, $B \ll B_{c2}$, the distance between vortices is larger than λ. Such vortices can be treated as independent non-interacting objects (see Section 2.2). In this limit, the collective effects can be ignored and the pinning force

7 Pinning investigation in NbN thin films

per vortex can be calculated as F_p^g/N, where F_p^g is the average pinning force and N is the number of vortices within the sample surface.

For the transport measurements the 100 μm wide bridge was structured by optical lithography and ion-beam etching. The pinning force is equal to the maximal sustainable Lorentz force that does not move vortices while current flows [Tin96]:

$$F_p^g(B) = VJ_cB = SdJ_cB, \qquad (7.6)$$

where J_c is the critical current density, S is the surface area of the sample and d is the sample thickness. The number of vortices can be calculated as:

$$N = BS/\phi_0. \qquad (7.7)$$

Hence the pinning force per vortex is given by:

$$F_p^g/N = J_c d\phi_0. \qquad (7.8)$$

The resulting temperature dependence of the pinning force per vortex F_p^g/N is shown in Table 7.2.

Tab. 7.2: Pinning force per vortex at different temperatures evaluated from the transport measurements for $\mu_0 H \approx 0$

Temperature (K)	8	10	12	14
J_c (10^5 A/cm^2)	10	5	3	2
F_p^g/N (pN)	2.07	1.04	0.62	0.42

From the MFM imaging the local force that the MFM tip exerts onto the vortices was evaluated to be about 0.74 pN (equ. 7.5). It was established, that this force leads to the depinning of vortices starting from 10 K. At 12 K this force overcomes the pinning force for almost all vortices. Consequently, the tip-vortex interaction force is equal to the mean local pinning force in the temperature range from 10 K to 12 K. This local value of the pinning force agrees very well with the global pinning force, F_p^g/N, evaluated from the transport measurements (see Table 7.2).

7.3 Global estimation of the pinning force

To conclude, the good agreement of the results estimated from local depinning of individual vortices by the MFM tip with the global transport measurements demonstrates that, for low fields $B \ll B_{c2}$, MFM is a powerful and reliable method to probe the local space variation of the pinning landscape. Additionally, it was demonstrated that the monopole-monopole model, despite it was derived for $d > 4\lambda$ [Car00], can be successfully applied even for thin films with a thickness $d \leq \lambda$.

8 Niobium/Permalloy hybrid structure: pinning by magnetic dots

The following chapter presents a local study of the vortex distribution in a conventional superconducting film in the vicinity of an artificially ordered pinning potential created by an array of ferromagnetic dots in a *magnetic-vortex* state. A direct correlation of the superconducting flux lines with the location of the magnetic dots is performed. The depinning of individual vortices by the stray field of the MFM tip additionally provides a possibility to estimate the pinning potential at different locations: on the top of ferromagnetic dots and in the interstitial positions.

Controlling the distribution of magnetic flux quanta (superconducting vortices) in superconducting materials by introducing artificial pinning centers is a challenge, both in basic and in applied research. Due to the presence of the natural point disorder (e.g. grain and intergrain pinning) in superconducting thin films superconducting vortices form a weakly disordered Abrikosov lattice [Vol02], a so-called topologically ordered Bragg glass [Gia97]. In the last decade a variety of studies has been performed to investigate the influence of different artificial pinning centers on the superconducting properties of thin films [Mos96, Bez96, Met99, Mos98b, Rei98, Rei97, Fie02, Bae95, Jia04, Zhu03, Mar97, Mor98, Bae01, Vil07, Jac98, Erd02, Bae03, Sil06, Sil07]. On the one hand, randomly distributed defects act as strong local pinning centers which significantly improve the in-field critical parameters of superconducting films [Fol07], on the other hand, ordered pinning potentials give rise to collective pinning mechanisms and thus lead to commensurate pinning effects [Mos96, Bez96, Met99, Mos98b, Rei98, Rei97, Fie02, Bae95, Jia04, Zhu03, Pat07]. In comparison to simple structurally ordered pinning sites, magnetic pinning centers provide additional degrees of freedom, which lead to several pronounced effects, such as domain-wall superconductivity, field induced superconductivity, proximity effect, magnetostatic interaction, and local suppression of superconductivity by strong out-of-plane field components [Mar97, Mor98, Bae01, Vil07, Jac98, Erd02, Bae03, Sil06, Sil07], some of which can be used to tune superconducting vortex dynamics by rectifying vortex motion (see for example [dSS07, Car05]). Thus, the investigation of the pinning mechanism in ferro-

8 Nb/Py hybrid structure: pinning by magnetic dots

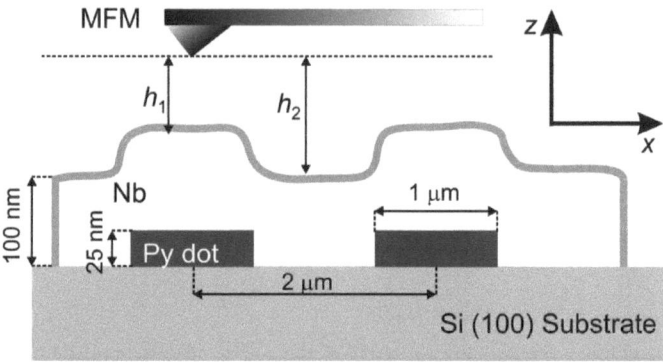

Fig. 8.1: Sketch of a cross-sectional cut of the FM/SC hybrid structure and the MFM tip scanning above the surface.

magnetic/superconducting (FM/SC) hybrid structures with well controlled lateral dimensions is of large scientific interest both in basic and in applied research.

8.1 Array of Permalloy dots

In this chapter the following hybrid structure is studied: a square array of permalloy (Py = $Ni_{80}Fe_{20}$) dots with 1 μm diameter, 25 nm height and 2 μm periodicity was prepared by Metlushko on Si (100) substrates using standard e-beam lithography, e-beam evaporation, and lift-off processes; a 100-nm-thick superconducting niobium (Nb) film (T_c = 8.32 K) was deposited on top of the Py dot array by sputter deposition [Hof08]. Figure 8.1 sketches the cross-sectional cut of this hybrid structure and shows the MFM tip scanning above the surface. A SEM scan of this Py/Nb hybrid structure is shown in Fig. 8.2 (inset).

Depending on their shape and aspect ratio, ferromagnetic dots can be in different magnetization states such as multidomain, single domain or, for the circular dots, a magnetic-vortex state becomes energetically stable at remanence [Sch00,Shi00]. Here, the magnetization curls continuously around the center while staying purely in-plane in a large area of the dot and turns perpendicular to the surface in the center of the dot creating a small magnetization swirl [Wac02,Hub09]. This swirl, also called a magnetic-vortex core, has either positive or negative polarity of the out-of-plane stray field and has a maximum width of $5l_{ex} \approx 25$ nm for the present geometry, with $l_{ex} = \sqrt{A/K_d}$ being the exchange length, where A is a material-specific exchange stiffness constant and K_d is the stray-field energy constant [Hub09].

The magnetic in-plane hysteresis loop (Fig. 8.2) of the Py array measured at 5 K using

8.1 Array of Permalloy dots

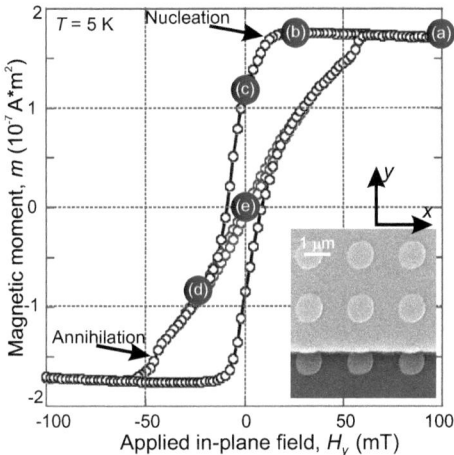

Fig. 8.2: Magnetic hysteresis loop, showing a magnetic vortex behavior with vortex nucleation and annihilation fields. Within the inner loop, between $-30\,\text{mT}$ and $+30\,\text{mT}$, the magnetization is reversible (vortex branch). The inset shows a SEM image of Py dots covered by a 100-nm thick Nb layer. The marked points on the loop correspond to the panels (a)-(e) in Fig. 8.3, respectively.

a superconducting quantum interference device (SQUID) clearly reveals a magnetic vortex behavior with vortex nucleation and annihilation fields. For the inner loop in the field range from $-30\,\text{mT}$ to $+30\,\text{mT}$, the magnetization process occurs only by vortex propagation and, thus, is reversible (vortex branch).

To reach the magnetic-vortex configuration in the Py array, the sample was cooled in the microscope to 40 K. An in-plane field of $+100\,\text{mT}$ was applied along the positive y direction (H_y) to ensure full saturation of the dots. Then the sample was cooled to 14.6 K. The MFM scan at a tip-sample distance of 75 nm shows four saturated Py dots (Fig. 8.3 (a)). Decreasing the field to $-25\,\text{mT}$ ensures that in most of the dots a magnetic vortex is nucleated ((Fig. 8.3 (c)). Going back to zero along the vortex branch brings the dots into the symmetric magnetic-vortex state imaged in Fig. 8.3 (e) and (e'). Magnetic vortices, generated in such a way, will have random polarity (i.e. out-of-plane magnetization components pointing randomly up or down) [Vil08]. To set a defined polarity, a small positive out-of-plane field ($+10\,\text{mT}$) was applied to the sample during the above-described field sequence. Thus, the magnetic-vortex core and the MFM tip, which is magnetized in positive z-direction, experience an attractive interaction that shows up as a dark contrast in the center of the dot (Fig. 8.3 (e')).

The measured shift of the resonant frequency Δf is proportional to the second derivative of

8 Nb/Py hybrid structure: pinning by magnetic dots

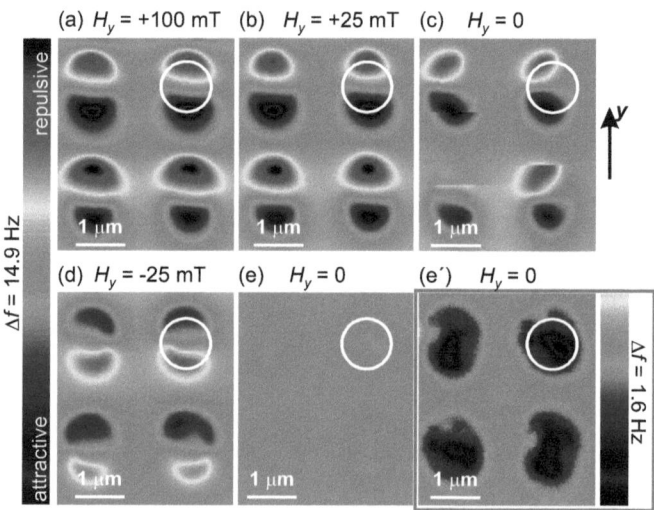

Fig. 8.3: To reach the magnetic vortex state the in-plane field H_y was varied along the hysteresis loop starting from saturation at (a) 100 mT, (b) 25 mT, (c) 0, (d) through applying a negative field less than the magnetic-vortex annihilation field (-25 mT) to the magnetic-vortex state at 0 mT (e) and (e'). Color bars give the measured Δf signal. A small out-of-plane field of 10 mT was permanently applied to set the orientation of the magnetic vortex (attractive interaction). The scanning area was 4 μm $\times$ 4 μm, the scanning distance 75 nm, $T = 14.6$ K. The white circle represents the location of the Py dot.

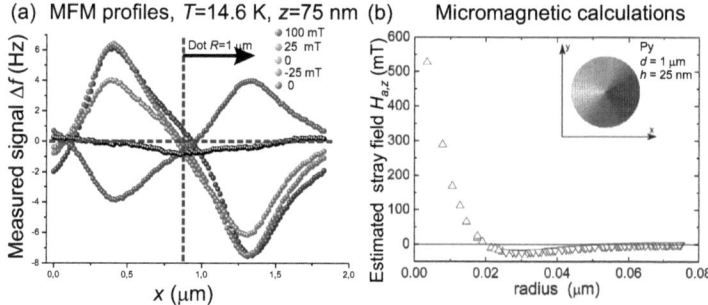

Fig. 8.4: (a) Line profiles taken from MFM images at different in-plane fields along the line going in the y direction through the center of the Py dot marked by the white circle (Fig. 8.3). The signal is proportional to the second derivative of the z component of the magnetic stray field of the Py dot. (b) Stray field distribution just above the surface of the Py dot in the vortex state. The estimated values are based on micromagnetic calculations using the LLC-program [Sch06].

the z component of the magnetic stray field of the sample on the set scanning distance from the surface (for details see Chapter 4). Fig. 8.4 (a) represents five line profiles taken from the MFM scans (Fig. 8.3) as lines along the y direction through the center of the marked dot (white circle) at different fields. In the in-plane saturated state the magnetic field distribution has the maximum out-of-plane component at the corresponding edges of the dot and zero value in the middle of the dot (Fig. 8.4 (a)). In the magnetic-vortex state the overall stray field above the dot is strongly reduced. As a result, the signal measured by MFM (black line, Fig. 8.4 (a)) for the magnetic-vortex state is almost 10 times less than for the saturated ones. This is also visible from the comparison of the scale bars in Fig. 8.3 (a) and (e'). At the same time a strong enhancement of the out-of-plane stray field component is expected at the center of the dot. Due to the fact that the 100 nm thick Nb film covers the Py array, imaging took place far away (about 175 nm) from the surface of the Py dot. Thus, the magnetic-vortex core is visualized by MFM as a slightly visible blue (attractive interaction) contrast. The stray field distribution just above the surface of a Py dot in the magnetic-vortex state is shown in Fig. 8.4 (b). The estimated values are based on micromagnetic calculations using the LLC-program [Sch06].

8.2 Low temperature experiments and discussion of the pinning mechanism

After reaching the magnetic-vortex state of the Py dots, as described above, the sample was cooled to a temperature below T_c of the Nb film ($T = 6.1$ K $= 72\%$ T_c) in perpendicular fields $H_{\text{applied}} = 0.5$ mT, 0, -0.5 mT and -1 mT, that are close to the matching fields for this hybrid structure. The matching field H_m is a field that ensures an integer number m of vortices per unit area (S) of the dot array: $H_m = m\phi_0/S$ [Mos96].

Hoffmann et al. have reported a clear correlation between a strong drop in the resistivity curve for a superconducting Nb film and the magnetic-vortex state of the underlying Py dots, which is shown to be independent on the polarity of the magnetic-vortex core [Hof08]. In the present work, local imaging was applied to look deeper into the nature of this enhanced pinning, providing a direct insight into the determination of the preferable locations of superconducting vortices, as well as to an estimation of the local pinning force by the direct depinning of individual flux lines.

An area where one dot is not fully switched to the magnetic-vortex state and has a residual in-plane component was chosen for LT-MFM imaging to confirm that the same dots are imaged at different fields and to correct a small thermal drift. It was established that the

8 Nb/Py hybrid structure: pinning by magnetic dots

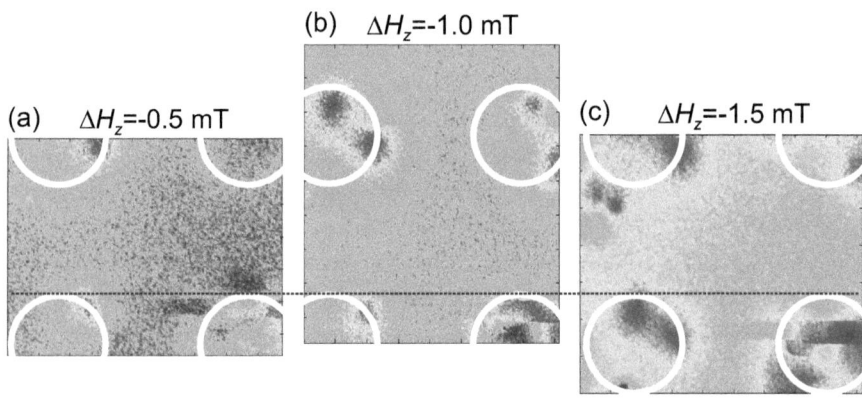

Fig. 8.5: Visualization of superconducting vortices pinned by Py dots at 6.1 K (72%T_c). For a better visualization of the vortex positions the "background" image was subtracted. The frozen fields ΔH_z are: (a) -0.5 mT, (b) -1 mT and (c) -1.5 mT. Superconducting vortices are visualized as red spots. The white circles represent the locations of four Py dots.

vertical coil of the microscope has a shift of its zero point in the range of 0.5 mT. This justifies to consider the 0.5 mT image, where only the magnetic contrast from the Py dots is observed, as a "background", and to subtract it from the other ones. The results are shown in Fig. 8.5 (a)–(c), respectively, and correspond to the reference fields $\Delta H_z = H_{\text{applied}} - 0.5$ mT. The orientation of ΔH_z is negative, so that the superconducting vortices and the MFM tip exhibit a repulsive interaction (red color). Hence, the superconducting vortices in the Nb film have an opposite polarity to that of the Py dots. This means that the magnetostatic interaction between magnetic and superconducting vortices is repulsive. Such a configuration is selected to differentiate the magnetostatic pinning mechanism from the non-magnetic one.

Figure 8.5 (a) corresponds to the first matching field H_1. Here one superconducting vortex, as expected, is visualized per unit area of the Py dot array. The superconducting vortices are located on top of the dots (white circles), showing that the Py dots work as preferable pinning centers. Nevertheless they do not concentrate at the center of the dot, but occupy the edges of the dot. Furthermore no superconducting vortices are visualized in the interstitial positions. This effect becomes more pronounced in Fig. 8.5 (b) where the second matching field H_2 has been applied during cooling. Also here, despite the long-range repulsive interaction between superconducting vortices, they are not distributed homogeneously, but are strongly pinned by each Py dot, so that two vortices are located on each dot.

A further increase of the field to H_3 leads to an enhanced magnetic contrast on top of the Py dots, which corresponds to multiple flux quanta pinned by the dots (Fig. 8.5 (c)).

8.2 Low temperature experiments and discussion of the pinning mechanism

Here the expected three superconducting vortices could not be separately resolved due to an overlapping of their magnetic lines at small vortex-vortex distances. The dark-blue contrast that appears at the edge of the lower left Py dot might be explained by the local polarization of the Py dot in the stray field coming out of the agglomeration of three superconducting vortices. This dark-blue contrast is stable and exists even at temperatures slightly above T_c of the Nb film (image is not shown here).

Estimation of the pinning force. A first rough estimation of the pinning force can be performed based on the fact that two superconducting vortices are observed to be situated close to each other on top of the Py dot rather than organized in a homogeneous Abrikosov lattice. Consequently, the pinning force at these artificial defects (F_p) is higher than the repulsive force between vortices $F_{\text{v-v}}$. The repulsive force between two SC vortices in thin films with a thickness below the penetration length λ is described as:

$$F_{\text{v-v}} = \frac{\phi_0^2}{\pi \mu_0 a^2}, \tag{8.1}$$

where a is the distance between the vortices [Pea64]. For the second matching field the distance between two vortices in a Nb film pinned by one Py dot was measured to be about $a = 750\,\text{nm}$ (Fig. 8.5 (b)). Thus, the vortex-vortex repulsion force normalized by the Nb film thickness was estimated to be $F_{\text{v-v}} \approx 19 \pm 0.3\,\text{pN}/\mu\text{m}$. According to this long-range interaction force, the formation of superconducting-vortex clusters in thin films is energetically unfavorable [Pea64]. Consequently, the presence of a strong pinning potential is required to ensure the distribution of superconducting vortices as visualized in Fig. 8.5 (b)–(c).

While scanning with the MFM tip, an additional force that acts on the superconducting vortices arises (see Chapter 4). This local tip-vortex interaction force can lead to a depinning of superconducting vortices and can be estimated from the MFM scans using the monopole-monopole model described elsewhere [Aus09, Str08]. From this model, the maximal lateral force that acts on the superconducting vortex from the MFM tip (see equ. 7.3, 7.4) increases proportionally to $(z + \lambda + d)^{-2}$ as the tip-sample distance z decreases.

Figure 8.6 shows MFM scans performed on the same sample position at different tip-sample distances in order to depin the superconducting vortices located on the Py dots and in the interstitial positions under the influence of the stray field of the tip. As long as the tip-sample distance h_2 (Fig. 8.1) is larger than 90 nm, the vortices are not dragged by the tip. As soon as h_2 reaches 90 nm (Fig. 8.6 (a)), the interstitial vortex (marked by the arrow) is depinned and moved completely away from the scanned area. This is apparent from the second scan at the

8 Nb/Py hybrid structure: pinning by magnetic dots

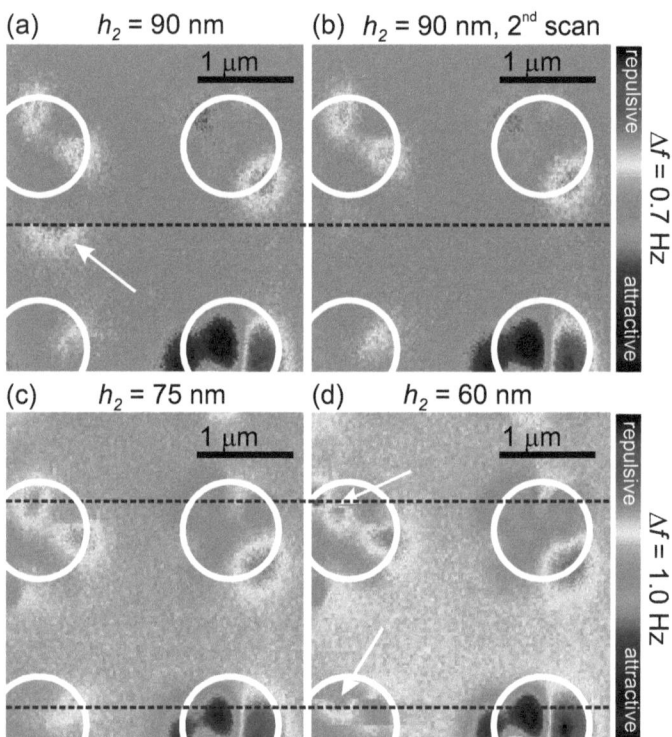

Fig. 8.6: Distribution of superconducting vortices in a Nb film in the presence of Py dots in the magnetic-vortex state measured at $T = 6.1\,\text{K}$ and $\Delta H_z = -1\,\text{mT}$. The superconducting vortices and the MFM tip exhibit an attractive interaction. The distance between the tip and the surface (h_2) decreases from panel (a), $h_2 = 90\,\text{nm}$, to panel (d), $h_2 = 60\,\text{nm}$. White circles represent the positions of the Py dots. The arrows mark the depinned vortices

8.2 Low temperature experiments and discussion of the pinning mechanism

same distance (Fig. 8.6 (b)), where no vortices are visible between the Py dots. The estimated pinning force for this interstitial superconducting vortex normalized by the Nb thickness is about $1.5\,\text{pN}/\mu\text{m}$. This value is 10 times less than the normalized pinning force $15\text{--}40\,\text{pN}/\mu\text{m}$ determined using the same model for the 300 nm thick film of pure Nb at 5.5 K [Str08]. The reason for this discrepancy may lie in the quality of the Nb films which is described by the number of natural defects. Another possible reason is a strong decrease of the pinning force when the film thickness becomes comparable to the penetration depth, $d \leq \lambda$.

The presence of the 25-nm-thick Py dots underneath the Nb film leads to a surface modulation of the superconducting film, as it is sketched in Fig. 8.1. The AFM profile (image not presented here) shows that the modulation $h_2 - h_1 \approx 30\,\text{nm}$ is on the scale of the Py thickness. Consequently, the superconducting vortices imaged on top of the Py dots have a lower tip-sample distance ($h_1 = 60\,\text{nm}$) and experience a stronger lateral force from the MFM tip. Despite the decreased distance, the vortices on the Py dots are not dragged by the tip at $h_2 = 90\,\text{nm}$. Only when h_2 decreases to 60 nm ($h_1 = 30\,\text{nm}$) and the depinning lateral force that acts in additional to the existing repulsive interaction (equ. 8.1) reaches $2.3\,\text{pN}/\mu\text{m}$, the vortices on top of the Py dots also start to move (Fig. 8.6 (d)). As a result, the total pinning force at the Py dots is estimated to be about $21\,\text{pN}/\mu\text{m}$. This value is about 15 times higher than the pinning force in pure Nb estimated above for the vortices in interstitial positions.

On the one hand, these microscopic observations support the conclusion made from the magnetoresistance measurement that the Py dots in the magnetic-vortex state act as highly preferable pinning sites [Hof08], on the other hand they show that another detailed explanation of the pinning mechanism is essential for understanding the visualized arrangement of superconducting vortices in FM/SC hybrid structures.

Two different mechanisms were proposed until now for the explanation of the enhanced pinning of superconducting vortices in the presence of a magnetic vortex. In the magnetostatic scenario, as it is described for example for an Al/Co hybrid structure [Vil08], pinning occurs due to the magnetostatic interaction of the superconducting vortex with the magnetic vortex. In such a case the superconducting vortices with opposite polarity to the magnetic vortex are expected to order themselves in interstitial positions of the dot array and the superconducting vortices with the same polarity have to be located directly at the magnetic vortex core. The preferred arrangement of superconducting vortices in interstitial positions can definitely be excluded from the present measurements. Another mechanism is based on the local suppression of superconductivity due to the highly localized out-of-plane field produced by the magnetic-vortex core (core pinning) [Hof08]. Here the magnetostatic interaction is negligible, and the superconducting vortices are located at the magnetic vortex core independently of their polarity. Nevertheless, neither of these mechanisms could fully explain the situation represented in the present work.

To exclude the possibility that the SC vortices are magnetostatically attracted by the returning stray field lines of the magnetic vortex, the stray field distribution just above the surface of a Py dot in the magnetic-vortex state was estimated by micromagnetic calculations [Sch06]. The estimated stray field for the Py dot geometry used in the above-described experiments reveals that the field lines return already at a distance of 25 nm from the center of the dot (Fig. 8.4 (b)). Hence, the simple attractive magnetostatic interaction between the SC vortex and the returning stray field of the magnetic vortex could not provoke the visualized arrangement of the SC vortices being located almost at the edges of the Py dot with $1\,\mu$m diameter (Fig. 8.5 (b)).

In summary, the results presented in this chapter demonstrate for the first time microscopically that the presence of magnetic dots in the magnetic-vortex state of the above deposited superconducting film significantly influences the natural pinning landscape. The superconducting vortices are preferably located at the edge of the Py dots, undergoing an about 15 times stronger pinning force at the Py dots compared to the pure Nb film. This pinning force overcomes the repulsive interaction between the SC vortices, allowing a superconducting vortex clusters to be pinned by a single dot. This local magnetic force microscopy study of the superconducting vortex distribution in the presence of an array of ferromagnetic dots with well controlled lateral dimensions is consistent with the global magnetoresistance measurements. Nevertheless, the observed local arrangement of the superconducting vortices could not be fully interpreted by the existing scenarios. Further local studies on such hybrid structures are of high scientific interest.

9 Conclusions

Controlling the distribution of magnetic flux quanta (superconducting vortices) in superconducting materials by introducing artificial pinning centers is a challenge, both in basic and in applied research. In this thesis the visualization of vortices in low- and high- temperature superconducting thin films and their pinning by natural and artificial defects have been reported.

For this purpose, low temperature magnetic force microscopy (LT-MFM) being the best compromise for vortex visualization and their correlation to topography in thin films was chosen. Revealing a lateral resolution of about 50 nm, LT-MFM is suitable to combine non-invasive imaging of separated vortices and their direct manipulation by the force that the MFM tip exerts on the individual vortex.

Since the surface roughness (droplets, precipitates) causes a severe problem to the scanning MFM tip, a nanoscale wedge polishing technique that allows to overcome this problem was developed (Chapter 6). First of all, it is shown that upon polishing the roughness of a sample can be reduced to a few nanometers. Additionally, not only the planarity of the film can be improved but also the polishing angle can be varied up to 10 angle-seconds. The developed nanoscale polishing is an alternative approach to investigate the internal structure of thin films and provides a continuous insight from the film surface down to the substrate. The main challenge of this work is to polish a wedge of a very thin film (200 nm) with an angle of a few seconds. As an example, this method was successfully used for the microstructural investigation of high temperature superconducting films $YBa_2Cu_3O_7$ (YBCO) with non-superconducting $BaHfO_3$ inclusions. It was demonstrated that, after polishing, the surface of even very rough chemical solution deposited films was smooth enough for a detailed high resolution electron backscatter diffraction (HR EBSD) analysis as well as for scanning force microscopy (SFM). The direct observation of artificial non-superconducting defects inside the YBCO film has been successfully performed and the homogeneous distribution of nanoparticles was confirmed by HR EBSD. Local vortex imaging on the polished films shows the unique application of this nanoscale polishing for local SFM studies on technically relevant films that have a high as-prepared roughness. Additionally, it was shown that the nanoscale

9 Conclusions

polishing is an important tool allowing to improve the surface quality of different kinds of thin films. The precipitates, droplets and surface corrugations that appear during the deposition process on the surface of the films were successfully removed. An interesting demonstration of the possibilities of the polishing technique was the creation of a unique topology in hard magnetic/superconducting heterostructures (Section 6.4).

To study the vortex distribution, its temperature dependence and a local estimation of the pinning force a 100 nm thin pulsed laser deposited NbN film was used (Chapter 7). When the temperature does not exceed 50% T_c, a non-invasive imaging of vortices is possible, indicating that the pinning force is larger than the dragging force of the MFM tip. Due to the presence of natural point disorder (e.g. grain and intergrain pinning) in superconducting thin films, superconducting vortices form a weakly disordered Abrikosov lattice, a so-called topologically ordered Bragg glass. Increasing the temperature leads to the partial movement of vortices to neighbouring positions revealing an inhomogeneous pinning potential. When the temperature reaches 75% T_c the pinning force for almost all vortices becomes equal to or less than the maximum lateral force that the MFM tip exerts on the vortex. As a result, such vortices are dragged by the tip and are imaged as "half-cut" objects. The pinning force estimated from this images using the monopole-monopole model for the tip-vortex interaction is (0.74 ± 0.04) pN. The good agreement of this local value with the pinning force calculated from global transport measurements demonstrates that for low fields, $B \ll B_{c2}$, MFM is a powerful and reliable method to probe the local variation of the pinning landscape.

Pinning is one of the most important properties of the superconductors on their way to industrial applications. Thus, the microscopical understanding of vortex pinning is indispensable for a future improvement of the performance of thin YBCO films, nowadays the most promising superconducting material for high-field and high-current applications. The main challenge concerning local studies (Chapter 5) is the possibility of a direct correlation between vortices and different kinds of artificial defects in YBCO films as well as vortex imaging on rough as-prepared thin films. For this purpose vortex imaging was performed at different frozen fields and temperatures on YBCO thin films modified with artificial defects of different size and composition (e.g. surface decoration and heavy ion irradiation). The results show that it is already challenging to visualize flux lines on the surface of YBCO films because of their high roughness. Another problem of such inhomogeneous films is that the local imaging of vortices gives no representative result for the whole film. However, the main feature that makes a correlation almost impossible is to find traces of nanodefects on the surface of the films, either within the LT scanning force microscope or with other external characterization techniques (e.g. HR EBSD), owing to the small-scaled nature of the defects.

Still, the surface-sensitive techniques and nanomanipulation are further improving, so that the results of this study suggest to approach this important topic with well pre-characterized surfaces for identifying the microstructural properties and a precisely located LT-MFM study on such an area, e.g. by using the designed search mask.

While randomly distributed defects act as strong local pinning centers which significantly improve the in-field critical parameters of the superconducting films, ordered pinning potentials give rise to collective pinning mechanisms and thus lead to commensurate pinning effects. To study the commensurate pinning effect locally, a ferromagnetic/superconducting hybrid structure constructed from a square array of circular 1-μm-sized ferromagnetic dots (Py = $Ni_{80}Fe_{20}$) covered by a 100 nm thin superconducting Nb film was investigated. In this thesis, it was demonstrated for the first time microscopically that the presence of magnetic vortices underneath the superconducting film significantly influences the natural pinning landscape (Chapter 8). The superconducting vortices are preferably located on top of the Py dots, experiencing an about 15 times stronger pinning force at the Py dots compared to the pure Nb film. This pinning force overcomes the repulsive interaction between the superconducting vortices, allowing superconducting vortex clusters to be pinned by each dot. These local magnetic force microscopy studies are consistent with global magnetoresistance measurements. Nevertheless, the reported local arrangement of the superconducting vortices could not be fully interpreted by the existing scenarios and requires further microscopical studies on hybrid structures.

List of publications

Part of the work accomplished during the PhD was published in the following journals.

[1] T. Shapoval, V. Neu, U. Wolff, R. Hühne, J. Hänisch, E. Backen, B. Holzapfel, L. Schultz. Study of pinning mechanisms in YBCO thin films by means of magnetic force microscopy. *Physica C* **460–462**, 732–733 (2007)

[2] T. Shapoval, S. Engel, M. Gründlich, D. Meier, E. Backen, V. Neu, B. Holzapfel and L. Schultz. Nanoscale wedge polishing of superconducting thin films - an easy way to obtain depth dependent information by surface analysis techniques. *Superconductor Science and Technology* **21**, 105015–(6) (2008) (Featured article of the month)

[3] J.T. Park, D.S. Inosov, Ch. Niedermayer, G.L. Sun, D. Haug, N.B. Christensen, R. Dinnebier, A.V. Boris, A.J. Drew, L. Schulz, T. Shapoval, U. Wolff, V. Neu, Xiaoping Yang, C.T. Lin, B. Keimer, and V. Hinkov. Electronic phase separation in the slightly underdoped iron pnictide superconductor $Ba_{1-x}K_xFe_2As_2$. *Physical Review Letters* **102**, 117006–(4) (2009)

[4] T. Shapoval, V. Metlushko, M. Wolf, B. Holzapfel, V. Neu, and L. Schultz. Direct observation of superconducting vortex clusters pinned by a periodic array of magnetic dots in ferromagnetic/superconducting hybrid structures. *Physical Review B* (brief reports) **81**, 92505–(4) (2010)

[5] J. Engelmann, T. Shapoval, S. Haindl, L. Schultz, and B. Holzapfel. Growth and characterization of $NbN/SmCo_5$ bilayers: The influence of magnetic stray field on the upper critical field of the superconductor. *Journal of Physics: Conference Series* **234**, 12012–(5) (2010)

[6] T. Shapoval, V. Metlushko, M. Wolf, V. Neu, B. Holzapfel, and L. Schultz. Enhanced pinning of superconducting vortices at circular magnetic dots in the magneticvortex state. *Physica C* **407**, 867–870 (2010)

[7] D. S. Inosov, T. Shapoval, V. Neu, U. Wolff, J. S. White, S. Haindl, J. T. Park, D. L. Sun, C. T. Lin, E. M. Forgan, M. S. Viazovska, J. H. Kim, M. Laver, K. Nenkov, O. Khvostikova,

S. Kühnemann, and V. Hinkov. Symmetry and disorder of the vitreous vortex lattice in overdoped BaFe$_{2-x}$Co$_x$As$_2$: Indication for strong single-vortex pinning. *Physical Review B* **81**, 014513–(12) (2010)

[8] S. Haindl, T. Thersleff, T. Shapoval, Y. W. Lai, J. McCord, L. Schultz, and B. Holzapfel. Advanced Nb/FePt L10 hybrid thin films. *Superconductor Science and Technology* **24**, 024002–(4) (2011)

[9] T. Shapoval, H. Stopfel, S. Haindl, J. Engelmann, D. S. Inosov, B. Holzapfel, V. Neu, and L. Schultz. Quantitative assessment of pinning forces and the superconducting gap in NbN thin films from complementary magnetic force microscopy and transport measurements. *arXiv:1101.4128v1, Submitted to Physical Review B*

Invited talk at the University of Salerno

T. Shapoval, V. Neu, E. Backen, S. Engel, M. Sparing, D. Meier, M. Gründlich, I. Mönch, U. Wolff, R. Hühne, B. Holzapfel and L. Schultz. Study of vortex pinning mechanism in YBCO thin films by low-temperature magnetic force microscopy. Seminar SUPERMAT, Department of Physics, University of Salerno/Italy, 17th of May (2007)

Bibliography

[Abr57] A. A. Abrikosov. On the magnetic properties of superconductors of the second group. *Soviet Physics JETP* **5**, 1174–1182 (1957)

[Abr85] A. A. Abrikosov. *Citation classic* **48**, 18 (1985)

[Ain94] M. Aindow and M. Yeadon. The origins of growth spirals on laser-ablated YBa$_2$Cu$_3$O$_{7-\delta}$ thin films. *Philosophical Magazine Letters* **70**, 47–53 (1994)

[Alb91] T. R. Albrecht, P. Grutter, D. Horne and D. Rugar. Frequency modulation detection using high-Q cantilevers for enhanced force microscope sensitivity. *J. Appl. Phys.* **69**, 668–673 (1991)

[Ale57] P. B. Alers. Structure of the intermediate state in superconducting lead. *Physical Review* **105**, 104–108 (1957)

[Aus09] O. M. Auslaender, L. Luan, E. W. J. Straver, J. E. Hoffman, N. C. Koshnick, E. Zeldov, D. A. Bonn, R. X. Liang, W. N. Hardy and K. A. Moler. Mechanics of individual isolated vortices in a cuprate superconductor. *Nature Physics* **5**, 35–39 (2009)

[Ayt08] T. Aytug, M. Paranthaman, K. J. Leonard, K. Kim, A. O. Ijaduola, Y. Zhang, E. Tuncer, J. R. Thompson and D. K. Christen. Enhanced flux pinning and critical currents in YBa$_2$Cu$_3$O$_{7-\delta}$ films by nanoparticle surface decoration: Extension to coated conductor templates. *J. Appl. Phys.* **104**, 043906–(6) (2008)

[Bac06] E. Backen. Private communication. Unpublished Work (2006)

[Bac07] E. Backen, J. Hänisch, R. Hühne, K. Tscharntke, S. Engel, T. Thersleff, L. Schultz and B. Holzapfel. Improved pinning in YBCO based quasi-multilayers prepared by on- and off-axis pulsed laser deposition. *IEEE Trans. Appl. Supercond.* **17**, 3733–3736 (2007)

[Bae95] M. Baert, V. V. Metlushko, R. Jonckheere, V. V. Moshchalkov and Y. Bruynseraede. Composite flux-line lattices stabilized in superconducting films by a regular array of artificial defects. *Phys. Rev. Lett.* **74**, 3269–3272 (1995)

[Bae01] M. J. Van Bael, J. Bekaert, K. Temst, L. Van Look, V. V. Moshchalkov, Y. Bruynseraede, G. D. Howells, A. N. Grigorenko, S. J. Bending and G. Borghs. Local observation of field polarity dependent flux pinning by magnetic dipoles. *Phys. Rev. Lett.* **86**, 155–158 (2001)

Bibliography

[Bae03] M. J. Van Bael, M. Lange, S. Raedts, V. V. Moshchalkov, A. N. Grigorenko and S. J. Bending. Local visualization of asymmetric flux pinning by magnetic dots with perpendicular magnetization. *Phys. Rev. B* **68**, 014509–(4) (2003)

[Bar57] J. Bardeen, L. N. Cooper and J. R. Schrieffer. Theory of superconductivity. *Physical Review* **108**, 1175–1204 (1957)

[Bea62] C. P. Bean. Magnetization of hard superconductors. *Phys. Rev. Lett.* **8**, 250–253 (1962)

[Bea64] C. P. Bean. Magnetization of high-field superconductors. *Rev. Mod. Phys.* **36**, 31–39 (1964)

[Bed86] J. G. Bednorz and K. A. Müller. Possible high T_c superconductivity in the Ba-La-Cu-O system. *Z. Phys. B* **64**, 189–193 (1986)

[Bez96] A. Bezryadin, Yu. N. Ovchinnikov and B. Pannetier. Nucleation of vortices inside open and blind microholes. *Phys. Rev. B* **53**, 8553–8560 (1996)

[Bin82] G. Binning, H. Rohrer, C. Gerber and E. Weibel. Surface studies by scanning tunneling microscopy. *Phys. Rev. Lett.* **49**, 57–61 (1982)

[Bit31] F. Bitter. On inhomogeneities in the magnetization of ferromagnetic materials. *Physical Review* **38**, 1903–1905 (1931)

[Bla94] G. Blatter, M. V. Feigelman, V. B. Geshkenbein, A. I. Larkin and V. M. Vinokur. Vortices in high-temperature superconductors. *Rev. Mod. Phys.* **66**, 1125–1388 (1994)

[Bon93] J. E. Bonevich, K. Harada, T. Matsuda, H. Kasai, T. Yoshida, G. Pozzi and A. Tonomura. Electron holography observation of vortex lattices in a superconductor. *Phys. Rev. Lett.* **70**, 2952–2955 (1993)

[Bra96] E. H. Brandt. Superconductors of finite thickness in a perpendicular magnetic field: Strips and slabs. *Phys. Rev. B* **54**, 4246–4264 (1996)

[Bra02a] E. H. Brandt. Vortices in superconductors. *Physica C* **369**, 10–20 (2002)

[Bra02b] E. H. Brandt, J. Vanacken and V. V. Moshchalkov. Vortices in physics. *Physica C* **369**, 1–9 (2002)

[Bra05] E. H. Brandt. Ginzburg-Landau vortex lattice in superconductor films of finite thickness. *Phys. Rev. B* **71**, 014521–(12) (2005)

[Bra09a] E. H. Brandt. Vortices in superconductors: ideal lattice, pinning, and geometry effects. *Supercond. Sci. Technol.* **22**, 034019–(9) (2009)

[Bra09b] E. H. Brandt and S.-P. Zhou. Attractive vortices. *Physics* **2**, 22–(4) (2009)

[Buc04] W. Buckel and R. Kleiner. *Supraleitung*. Wiley-VCH, Weinheim (2004)

Bibliography

[Bur94] J. Burger, M. Lippert, W. Dorsch, P. Bauer and G. Saemann-Ischenko. Observation of the early stages of growth and the formation of growth spirals in epitaxial $YBa_2Cu_3O_{7-\delta}$ thin films by AFM. *Applied Physics A* **59**, 49–56 (1994)

[Car00] G. Carneiro and E. H. Brandt. Vortex lines in films: Fields and interactions. *Phys. Rev. B* **61**, 6370–6376 (2000)

[Car05] G. Carneiro. Tunable ratchet effects for vortices pinned by periodic magnetic dipole arrays. *Physica C* **432**, 206–214 (2005)

[Cas03] O. Castano, A. Cavallaro, A. Palau, J. C. Gonzalez, M. Rossell, T. Puig, F. Sandiumenge, N. Mestres, S. Pinol, A. Pomar and X. Obradors. High quality $YBa_2Cu_3O_7$ thin films grown by trifluoroacetates metalorganic deposition. *Supercond. Sci. Technol.* **16**, 45–53 (2003)

[Cha92] A. M. Chang, H. D. Hallen, L. Harriott, H. F. Hess, H. L. Kao, J. Kwo, R. E. Miller, R. Wolfe, J. van der Ziel and T. Y. Chang. Scanning Hall probe microscopy. *Appl. Phys. Lett.* **61**, 1974–1976 (1992)

[Cho01] B. K. Chong, H. Zhou, G. Mills, L. Donaldson and J. M. R. Weaver. Scanning Hall probe microscopy on an atomic force microscope tip. *J. Vac. Sci. Technol. A.* **19**, 1769–1772 (2001)

[Civ97] L. Civale. Vortex pinning and creep in high-temperature superconductors with columnar defects. *Supercond. Sci. Technol.* **10**, A11–A28 (1997)

[Cri64] D. Cribier, B. Jacrot, L. Madhav Rao and B. Farnoux. Mise en evidence par diffraction de neutrons d'une structure periodique du champ magnetique dans le niobium supraconducteur. *Phys. Letters* **9**, 106–107 (1964)

[Ded06] M. Dede, A. Oral, T. Yamamoto, K. Kadowaki and H. Shtrikman. Real-time imaging of vortex-antivortex annihilation in $Bi_2Sr_2CaCu_2O_{8+\delta}$ single crystals by low temperature scanning Hall probe microscopy. *Jpn. J. Appl. Phys.* **45**, 2246–2250 (2006)

[DeS60] W. DeSorbo. Study of the intermediate state in superconductors using cerium phosphate glass. *Phys. Rev. Lett.* **4**, 406–408 (1960)

[Dor92] L. A. Dorosinskii, M. V. Indenbom, V. I. Nikitenko, Yu. A. Ossip'yan, A. A. Polyanskii and V. K. Vlasko-vlasov. Studies of HTSC crystal magnetization features using indicator magnetooptic films with in-plane anisotropy. *Physica C* **203**, 149–156 (1992)

[dSS07] C. C. de Souza Silva, A. V. Silhanek, J. Van de Vondel, W. Gillijns, V. Metlushko, B. Ilic and V. V. Moshchalkov. Dipole-induced vortex ratchets in superconducting films with arrays of micromagnets. *Phys. Rev. Lett.* **98**, 117005–(4) (2007)

[Elr84] S. A. Elrod, A. L. de Lozanne and C. F. Quate. Low-temperature vacuum tunneling microscopy. *Appl. Phys. Lett.* **45**, 1240–1242 (1984)

Bibliography

[Eng07] S. Engel, T. Thersleff, R. Hühne, L. Schultz and B. Holzapfel. Enhanced flux pinning in $YBa_2Cu_3O_7$ layers by the formation of nanosized $BaHfO_3$ precipitates using the chemical deposition method. *Appl. Phys. Lett.* **90**, 102505–(3) (2007)

[Erd02] S. Erdin, A. F. Kayali, I. F. Lyuksyutov and V. L. Pokrovsky. Interaction of mesoscopic magnetic textures with superconductors. *Phys. Rev. B* **66**, 014414–(7) (2002)

[Ess67] U. Essmann and H. Träuble. The direct observation of individual flux lines in type II superconductors. *Phys. Lett.* **24A**, 526–527 (1967)

[Fal02] M. Falter, W. Hässler, B. Schlobach and B. Holzapfel. Chemical solution deposition of $YBa_2Cu_3O_{7-x}$ films by dip coating. *Physica C* **372-376**, 46–49 (2002)

[Fas06] Y. Fasano, M. De Seta, M. Menghini, H. Pastoriza and F. de la Cruz. Effect of vortex dimensionality on elastic and plastic symmetry transformations induced in vortex matter. *Journal of Physics and Chemistry of Solids* **67**, 395–398 (2006)

[Fas08] Y. Fasano and M. Menghini. Magnetic-decoration imaging of structural transitions induced in vortex matter. *Supercond. Sci. Technol.* **21**, 023001–(24) (2008)

[Fel05] D. M. Feldmann, D. C. Larbalestier, T. Holesinger, R. Feenstra, A. A. Gapud and E. D. Specht. Evidence for extensive grain boundary meander and overgrowth of substrate grain boundaries in high critical current density ex situ $YBa_2Cu_3O_{7-x}$ coated conductors. *Journal of Materials Research* **20**, 2012–2020 (2005)

[Fer03] L. Fernández, B. Holzapfel, F. Schindler, B. de Boer, A. Attenberger, J. Hänisch and L. Schultz. Influence of the grain boundary network on the critical current of $YBa_2Cu_3O_7$ films grown on biaxially textured metallic substrates. *Phys. Rev. B* **67**, 052503–(4) (2003)

[Fie02] S. B. Field, S. S. James, J. Barentine, V. Metlushko, G. Crabtree, H. Shtrikman, B. Ilic and S. R. J. Brueck. Vortex configurations, matching, and domain structure in large arrays of artificial pinning centers. *Phys. Rev. Lett.* **88**, 067003–(4) (2002)

[Fis07] O. Fischer, M. Kugler, I. Maggio-Aprile, C. Berthod and C. Renner. Scanning tunneling spectroscopy of high-temperature superconductors. *Rev. Mod. Phys.* **79**, 353–419 (2007)

[Fit66] W. Fite and A. G. Redfield. Superconducting mixed-state-structure determination in vanadium by nuclear magnetic resonance and relaxation. *Phys. Rev. Lett.* **17**, 381–383 (1966)

[Fol07] S. R. Foltyn, L. Civale, J. L. MacManus-Driscoll, Q. X. Jia, B. Maiorov, H. Wang and M. Maley. Materials science challenges for high-temperature superconducting wire. *Nature Materials* **6**, 631–642 (2007)

[Fuc06] G. Fuchs. Private communication. Unpublished Work (2006)

[Ger91] C. Gerber, D. Anselmetti, J. G. Bednorz, J. Mannhart and D. G. Schlom. Screw dislocations in high-T_c films. *Nature* **350**, 279–280 (1991)

[Gia97] T. Giamarchi and P. Le Doussal. Phase diagrams of flux lattices with disorder. *Phys. Rev. B* **55**, 6577–6583 (1997)

[Gib07] M. Gibert, T. Puig, X. Obradors, A. Benedetti, F. Sandiumenge and R. Hühne. Self-organization of heteroepitaxial CeO_2 nanodots grown from chemical solutions. *Adv. Mater.* **19**, 3937–3842 (2007)

[Gin50] V. L. Ginzburg and L. D. Landau. On the theory of superconductivity. *Pis'ma Zhurnal Eksperimental'noi i Teoreticheskoi Fiziki* **20**, 1064–1082 (1950)

[Goa01] P. E. Goa, H. Hauglin, M. Baziljevich, E. Il'yashenko, P. L. Gammel and T. H. Johansen. Real-time magneto-optical imaging of vortices in superconducting $NbSe_2$. *Supercond. Sci. Technol.* **14**, 729–731 (2001)

[Gor59] L. P. Gor'kov. Microscopic derivation of the Ginzburg-Landau equations in the theory of superconductivity. *Soviet Physics JETP* **9**, 1364–1367 (1959)

[Gri07] I. V. Grigorieva, W. Escoffier, V. R. Misko, B. J. Baelus, F. M. Peeters, L. Y. Vinnikov and S. V. Dubonos. Pinning-induced formation of vortex clusters and giant vortices in mesoscopic superconducting disks. *Phys. Rev. Lett.* **99**, 147003–(4) (2007)

[Gut07] J. Gutierrez, A. Llordes, J. Gazquez, M. Gibert, N. Roma, S. Ricart, A. Pomar, F. Sandiumenge, N. Mestres, T. Puig and X. Obradors. Strong isotropic flux pinning in solution-derived $YBa_2Cu_3O_{7-x}$ nanocomposite superconductor films. *Nature Materials* **6**, 367–373 (2007)

[Haa96] T. Haage, Q. D. Jiang, M. Cardona, H.-U. Habermeier and J. Zegenhagen. Direct scanning tunneling microscopy observation of non-unit-cell growth of $YBa_2Cu_3O_{7-\delta}$ thin films. *Appl. Phys. Lett.* **68**, 2427–2429 (1996)

[Hai08] S. Haindl, M. Weisheit, T. Thersleff, L. Schultz and B. Holzapfel. Enhanced field compensation effect in superconducting/hard magnetic Nb/FePt bilayers. *Supercond. Sci. Technol.* **21**, 045017–(6) (2008)

[Hai09] S. Haindl. In preparation. Unpublished Work (2009)

[Han83] P. Hansen, K. Witter and W. Tolksdorf. Magnetic and magneto-optic properties of lead-substituted and bismuth-substituted yttrium-iron-garnet films. *Phys. Rev. B* **27**, 6608–6625 (1983)

[Har92] K. Harada, T. Matsuda, J. Bonevich, M. Igarashi, S. Kondo, G. Pozzi, U. Kawabe and A. Tonomura. Real-time observation of vortex lattices in a superconductor by electron-microscopy. *Nature* **360**, 51–53 (1992)

[Has08] K. Hasselbach, C. Ladam, V. O. Dolocan, D. Hykel, T. Crozes, K. Schuster and D. Mailly. High resolution magnetic imaging: microSQUID force microscopy. *Journal of Physics* **97**, 012330–(6) (2008)

Bibliography

[Hau04] T. Haugan, P. N. Barnes, R. Wheeler, F. Meisenkothen and M. Sumption. Addition of nanoparticle dispersions to enhance flux pinning of the $YBa_2Cu_3O_{7-x}$ superconductor. *Nature* **430**, 867–870 (2004)

[Haw91] M. Hawley, I. D. Raistrick, J. G. Beery and R. J. Houlton. Growth-mechanism of sputtered films of $YBa_2Cu_3O_7$ studied by scanning tunneling microscopy. *Science* **251**, 1587–1589 (1991)

[Hes89] H. F. Hess, R. B. Robinson, R. C. Dynes, J. M. Valles and J. V. Waszczak. Scanning-tunneling-microscope observation of the Abrikosov flux lattice and the density of states near and inside a fluxoid. *Phys. Rev. Lett.* **62**, 214–216 (1989)

[Hän04] J. Hänisch. Private communication. Unpublished Work (2004)

[Hän05] J. Hänisch, C. Cai, R. Hühne, L. Schultz and B. Holzapfel. Formation of nanosized $BaIrO_3$ precipitates and their contribution to flux pinning in Ir-doped $YBa_2Cu_3O_{7-\delta}$ quasi-multilayers. *Appl. Phys. Lett.* **86**, 122508–(3) (2005)

[Hän06] J. Hänisch, C. Cai, V. Stehr, R. Hühne, J. Lyubina, K. Nenkov, G. Fuchs, L. Schultz and B. Holzapfel. Formation and pinning properties of growth-controlled nanoscale precipitates in $YBa_2Cu_3O_{7-\delta}$/transition metal quasi-multilayers. *Supercond. Sci. Technol.* **19**, 534–540 (2006)

[Hof08] A. Hoffmann, L. Fumagalli, N. Jahedi, J. C. Sautner, J. E. Pearson, G. Mihajlovic and V. Metlushko. Enhanced pinning of superconducting vortices by magnetic vortices. *Phys. Rev. B* **77**, 060506(R)–(4) (2008)

[Hol92] B. Holzapfel, B. Roas, L. Schultz, P. Bauer and G. Saemann-Ischenko. Off-axis laser deposition of $YBa_2Cu_3O_{7-\delta}$ thin films. *Appl. Phys. Lett.* **61**, 3178–3180 (1992)

[Hol93] B. Holzapfel, G. Kreiselmeyer, M. Kraus, G. Saemann-Ischenko, S. Bouffard, S. Klaumünzer and L. Schultz. Angle-resolved critical transport-current density of $YBa_2Cu_3O_{7-\delta}$ thin-films and $YBa_2Cu_3O_{7-\delta}$/$PrBa_2Cu_3O_{7-\delta}$ superlattices containing columnar defects of various orientations. *Phys. Rev. B* **48**, 600–603 (1993)

[Hor07] I. Horcas, R. Fernandez, J. M. Gomez-Rodriguez, J. Colchero, J. Gomez-Herrero and A. M. Baro. WSXM: A software for scanning probe microscopy and a tool for nanotechnology. *Rev. Sci. Instrum.* **78**, 013705–(8) (2007)

[Hor08] T. Horide, K. Matsumoto, P. Mele, A. Ichinose, R. Kita, M. Mukaida, Y. Yoshida and S. Horii. The crossover from the vortex glass to the Bose glass in nanostructured $YBa_2Cu_3O_{7-x}$ films. *Appl. Phys. Lett.* **92**, 182511–(3) (2008)

[Hub09] A. Hubert and R. Schäfer. *Magnetic domains*. Springer Berlin (2009)

[Hug98] H. J. Hug, B. Stiefel, P. J. A. van Schendel, A. Moser, R. Hofer, S. Martin, H.-J. Güntherodt, S. Porthun, L. Abelmann, J. C. Lodder, G. Bochi and R. C. O'Handley. Quantitative magnetic force microscopy on perpendicularly magnetized samples. *J. Appl. Phys.* **83**, 5609–5620 (1998)

[Jac98] Y. Jaccard, J. I. Martin, M.-C. Cyrille, M. Velez, J. L. Vicent and I. K. Schuller. Magnetic pinning of the vortex lattice by arrays of submicrometric dots. *Phys. Rev. B* **58**, 8232–8235 (1998)

[Jia04] Z. Jiang, D. A. Dikin, V. Chandrasekhar, V. V. Metlushko and V. V. Moshchalkov. Pinning phenomena in a superconducting film with a square lattice of artificial pinning centers. *Appl. Phys. Lett.* **84**, 5371–5373 (2004)

[Joo02] Ch. Jooss, J. Albrecht, H. Kuhn, S. Leonhardt and H. Kronmüller. Magneto-optical studies of current distributions in high-T_c superconductors. *Reports on Progress in Physics* **65**, 651–788 (2002)

[Kam08] Y. Kamihara, T. Watanabe, M. Hirano and H. Hosono. Iron-based layered superconductor La[$O_{1-x}F_x$]FeAs (x =0.05-0.12) with T_c = 26 K. *J. Am. Chem. Soc.* **130**, 3296–3297 (2008)

[Kan07] S. Kang, K. J. Leonard, P. M. Martin, J. Li and A. Goyal. Strong enhancement of flux pinning in YBa$_2$Cu$_3$O$_{7-\delta}$ multilayers with columnar defects comprised of self-assembled BaZrO$_3$ nanodots. *Supercond. Sci. Technol.* **20**, 11–15 (2007)

[Kaw90] T. Kawasaki, T. Matsuda, J. Endo and A. Tonomura. Observation of a 0.055 nm spacing lattice image in gold using a field emission electron microscope. *Japanese Journal of Applied Physics* **29**, L508–L510 (1990)

[Kir96] J. R. Kirtley, C. C. Tsuei, M. Rupp, J. Z. Sun, L. S. Yu-Jahnes, A. Gupta, M. B. Ketchen, K. A. Moler and M. Bhushan. Direct imaging of integer and half-integer Josephson vortices in high-T_c grain boundaries. *Phys. Rev. Lett.* **76**, 1336–1339 (1996)

[Kon06] J. Konrad, S. Zaefferer and D. Raabe. Investigation of orientation gradients around a hard Laves particle in a warm-rolled Fe$_3$Al-based alloy using a 3D EBSD-FIB technique. *Acta Materialia* **54**, 1369–1380 (2006)

[Kra93] M. Kraus, P. Van Hasselt, J. P. Ströbel, S. Peehs, M. Leghissa, G. Kreiselmeyer, B. Holzapfel, W. Gerhäuser, B. Hensel, S. Klaumünzer, S. Bouffard and G. Saemann-Ischenko. Heavy-ion irradiation of high-temperature-superconductors: Material modifications and influence on pinning mechanisms. *Radiation Effects and Defects in Solids* **126**, 147–150 (1993)

[Kre07] H.-U. Krebs. *Pulsed laser deposition of thin metals. in R. Eason, Editor: Pulsed laser deposition of thin films.* Wiley (2007)

[Kub84] S. Kubo, M. Asahi, M. Hikita and M. Igarashi. Magnetic penetration depths in superconducting NbN films prepared by reactive dc magnetron sputtering. *Appl. Phys. Lett.* **44**, 258–260 (1984)

[Kur04] A. Kursumovic, R. I. Tomov, R. Hühne, J. L. MacManus-Driscoll, B. A. Glowacki and J. E. Evetts. Hybrid liquid phase epitaxy processes for YBa$_2$Cu$_3$O$_7$ film growth. *Supercond. Sci. Technol.* **17**, 1215–1223 (2004)

Bibliography

[Lan03] M. Lange, M. J. Van Bael, Y. Bruynseraede and V. V. Moshchalkov. Nanoengineered magnetic-field-induced superconductivity. *Phys. Rev. Lett.* **90**, 197006–(4) (2003)

[Loh00] J. Lohau, S. Kirsch, A. Carl and E. F. Wassermann. Quantitative determination of the magnetization and stray field of a single domain Co/Pt dot with magnetic force microscopy. *Appl. Phys. Lett.* **76**, 3094–3096 (2000)

[Lon35] F. London and H. London. The electromagnetic equations of the supraconductor. *Proc. Royal Society London A* **149**, 71–88 (1935)

[Loz99] A. De Lozanne. Scanning probe microscopy of high-temperature superconductors. *Supercond. Sci. Technol.* **12**, R43–R56 (1999)

[Lu02] Q. Lu, K. Mochizuki, J. T. Markert and A. De Lozanne. Localized measurement of penetration depth for a high T_c superconductor single crystal using a magnetic force microscope. *Physica C* **371**, 146–150 (2002)

[Luk08] A. Lukashenko, R. Wördenweber and A. V. Ustinov. Imaging of vortex flow in microstructured high-T_c films by laser scanning microscope. *Physica C* **468**, 552–556 (2008)

[MA95] I. Maggio-Aprile, Ch. Renner, A. Erb, E. Walker and O. Fischer. Direct vortex lattice imaging and tunneling spectroscopy of flux lines on $YBa_2Cu_3O_{7-\delta}$. *Phys. Rev. Lett.* **75**, 2754–2757 (1995)

[Mar97] J. I. Martin, M. Velez, J. Nogues and I. K. Schuller. Flux pinning in a superconductor by an array of submicrometer magnetic dots. *Phys. Rev. Lett.* **79**, 1929–1932 (1997)

[Mat89] T. Matsuda, S. Hasegawa, M. Igarashi, T. Kobayashi, M. Naito, H. Kajiyama, J. Endo, N. Osakabe, A. Tonomura and R. Aoki. Magnetic-field observation of a single flux quantum by electron-holographic interferometry. *Phys. Rev. Lett.* **62**, 2519–2522 (1989)

[Mat04] K. Matsumoto, T. Horide, K. Osamura, M. Mukaida, Y. Yoshida, A. Ichinose and S. Horii. Enhancement of critical current density of YBCO films by introduction of artificial pinning centers due to the distributed nano-scaled Y_2O_3 islands on substrates. *Physica C* **412-414**, 1267–1271 (2004)

[Maz98] A. Mazilu, H. Safar, M. P. Maley, J. Y. Coulter, L. N. Bulaevskii and S. Foltyn. Vortex dynamics of heavy-ion-irradiated $YBa_2Cu_3O_{7-\delta}$: Experimental evidence for a reduced vortex mobility at the matching field. *Phys. Rev. B* **58**, R8909–R8912 (1998)

[MD08] J. L. MacManus-Driscoll, P. Zerrer, H. Y. Wang, H. Yang, J. Yoon, A. Fouchet, R. Yu, M. G. Blamire and Q. X. Jia. Strain control and spontaneous phase ordering in vertical nanocomposite heteroepitaxial thin films. *Nature Materials* **7**, 314–320 (2008)

[Mei33] W. Meissner and R. Ochsenfeld. Ein neuer Effekt bei Eintritt der Supraleitfahigkeit. *Die Naturwissenschaften* **21**, 787 (1933)

[Mei09] D. Meier. Diploma thesis TU Dresden. Unpublished Work (2009)

[Met99] V. Metlushko, U. Welp, G. W. Crabtree, Z. Zhang, S. R. J. Brueck, B. Watkins, L. E. Delong, B. Ilic, K. Chung and P. J. Hesketh. Nonlinear flux-line dynamics in vanadium films with square lattices of submicron holes. *Phys. Rev. B* **59**, 603–607 (1999)

[Mön07] J. I. Mönch. Private communication. Unpublished Work (2007)

[Mor98] D. J. Morgan and J. B. Ketterson. Asymmetric flux pinning in a regular array of magnetic dipoles. *Phys. Rev. Lett.* **80**, 3614–3617 (1998)

[Mor09] C. Moreno, P. Abellán, A. Hassini, A. Ruyter, A. Perez del Pino, F. Sandiumenge, M.-J. Casanove, J. Santiso, T. Puig and X. Obradors. Spontaneous outcropping of self-assembled insulating nanodots in solution derived metallic ferromagnetic $La_{0.7}Sr_{0.3}MnO_3$ Films. *Advanced Functional Materials* (2009)

[Mos95] A. Moser, H. J. Hug, I. Parashikov, B. Stiefel, O. Fritz, H. Thomas, K. Baratoff, H.-J. Güntherodt and P. Chaudhari. Observation of single vortices condensed into a vortex-glass phase by magnetic force microscopy. *Phys. Rev. Lett.* **74**, 1847–1850 (1995)

[Mos96] V. V. Moshchalkov, M. Baert, V. V. Metlushko, E. Rosseel, M. J. Van Bael, K. Temst, R. Jonckheere and Y. Bruynseraede. Magnetization of multiple-quanta vortex lattices. *Phys. Rev. B* **54**, 7385–7393 (1996)

[Mos98a] A. Moser, H. J. Hug, B. Stiefel and H.-J. Güntherodt. Low temperature magnetic force microscopy on $YBa_2Cu_3O_{7-\delta}$ thin films. *Journal of Magnetism and Magnetic Materials* **190**, 114–123 (1998)

[Mos98b] V. V. Moshchalkov, M. Baert, V. V. Metlushko, E. Rosseel, M. J. Van Bael, K. Temst, Y. Bruynseraede and R. Jonckheere. Pinning by an antidot lattice: The problem of the optimum antidot size. *Phys. Rev. B* **57**, 3615–3622 (1998)

[Obr06] X. Obradors, T. Puig, A. Pomar, F. Sandiumenge, N. Mestres, M. Coll, A. Cavallaro, N. Roma, J. Gazquez, J. C. Gonzalez, O. Castano, J. Gutierrez, A. Palau, K. Zalamova, S. Morlens, A. Hassini, M. Gibert, S. Ricart, J. M. Moreto, S. Pinol, D. Isfort and J. Bock. Progress towards all-chemical superconducting $YBa_2Cu_3O_7$-coated conductors. *Supercond. Sci. Technol.* **19**, S13–S26 (2006)

[Onn11] H. K. Onnes. The superconductivity of mercury. *Comm. Phys. Lab. Univ. Leiden* **Nos. 122 and 124** (1911)

[Ora96] A. Oral, S. J. Bending and M. Henini. Real-time scanning Hall probe microscopy. *Appl. Phys. Lett.* **69**, 1324–1326 (1996)

Bibliography

[Ora97]   A. Oral, S. J. Bending, R. G. Humphreys and M. Henini. Microscopic measurement of penetration depth in YBa$_2$Cu$_3$O$_{7-\delta}$ thin films by scanning Hall probe microscopy. *Supercond. Sci. Technol.* **10**, 17–20 (1997)

[Ora98]   A. Oral, J. C. Barnard, S. J. Bending, I. I. Kaya, S. Ooi, T. Tamegai and M. Henini. Direct observation of melting of the vortex solid in Bi$_2$Sr$_2$CaCu$_2$O$_{8+\delta}$ single crystals. *Phys. Rev. Lett.* **80**, 3610–3613 (1998)

[Pat07]   U. Patel, Z. L. Xiao, J. Hua, T. Xu, D. Rosenmann, V. Novosad, J. Pearson, U. Welp, W. K. Kwok and G. W. Crabtree. Origin of the matching effect in a superconducting film with a hole array. *Phys. Rev. B* **76**, 020508(R)–(4) (2007)

[Pea64]   J. Pearl. Current distribution in superconducting films carrying quantized fluxoids. *Appl. Phys. Lett.* **5**, 65–66 (1964)

[Pi04]   U. H. Pi, Z. G. Khim, D. H. Kim, A. Schwarz, M. Liebmann and R. Wiesendanger. Dynamic force spectroscopy across an individual strongly pinned vortex in a Bi$_2$Sr$_2$CaCu$_2$O$_{8+\delta}$ single crystal. *Appl. Phys. Lett.* **85**, 5307–5309 (2004)

[Poo95]   P. Poole, H. A. Farach and R. J. Creswick. *Superconductivity*. Academic Press, San Diego, London (1995)

[Rei97]   C. Reichhardt, C. J. Olson and F. Nori. Dynamic phases of vortices in superconductors with periodic pinning. *Phys. Rev. Lett.* **78**, 2648–2651 (1997)

[Rei98]   C. Reichhardt, C. J. Olson and F. Nori. Commensurate and incommensurate vortex states in superconductors with periodic pinning arrays. *Phys. Rev. B* **57**, 7937–7943 (1998)

[Ros00]   M. Roseman and P. Grütter. Cryogenic magnetic force microscope. *Rev. Sci. Instrum.* **71**, 3782–3787 (2000)

[Ros01]   M. Roseman and P. Grütter. Estimating the magnetic penetration depth using constant-height magnetic force microscopy images of vortices. *New Journal of Physics* **3**, 24.1–24.8 (2001)

[Ros02]   M. Roseman and P. Grütter. Magnetic imaging and dissipation force microscopy of vortices on superconducting Nb films. *Applied Surface Science* **188**, 416–420 (2002)

[Rug89]   D. Rugar, H. J. Mamin and P. Guethner. Improved fiber-optic interferometer for atomic force microscopy. *Appl. Phys. Lett.* **55**, 2588–2590 (1989)

[San04]   A. Sandhu, K. Kurosawa, M. Dede and A. Oral. 50 nm Hall sensors for room temperature scanning Hall probe microscopy. *Japanese Journal of Applied Physics* **43**, 777–778 (2004)

[Sch94]   D. G. Schlom, D. Anselmetti, J. G. Bednorz, Ch. Gerber and J. Mannhart. Epitaxial-growth of cuprate superconductors from the gas phase. *Journal of Crystal Growth* **137**, 259–267 (1994)

Bibliography

[Sch00] M. Schneider, H. Hoffmann and J. Zweck. Lorentz microscopy of circular ferromagnetic permalloy nanodisks. *Appl. Phys. Lett.* **77**, 2909–2911 (2000)

[Sch06] M. R. Scheinfein. The LLG Micromagnetics Simulator$^{(TM)}$ http://llgmicro.home.mindspring.com. Unpublished Work (2006)

[Sha07] T. Shapoval, V. Neu, U. Wolff, R. Hühne, J. Hänisch, E. Backen, B. Holzapfel and L. Schultz. Study of pinning mechanisms in YBCO thin films by means of magnetic force microscopy. *Physica C* **460-462**, 732–733 (2007)

[Sha08] T. Shapoval, S. Engel, M. Gründlich, D. Meier, E. Backen, V. Neu, B. Holzapfel and L. Schultz. Nanoscale wedge polishing of superconducting thin films - an easy way to obtain depth dependent information by surface analysis techniques. *Supercond. Sci. Technol.* **21**, 105015–(6) (2008)

[Shi00] T. Shinjo, T. Okuno, R. Hassdorf, K. Shigeto and T. Ono. Magnetic vortex core observation in circular dots of permalloy. *Science* **289**, 930–932 (2000)

[Shi08] M. Shimizu, Y. Matsushima, M. Uno, K. Satoh, T. Yotsuya and T. Ishida. Little-Parks effect of fine superconducting Pb networks fabricated by focused-ion-beam microscope. *Physica C* **468**, 1298–1300 (2008)

[Sil06] A. V. Silhanek, W. Gillijns, V. V. Moshchalkov, V. Metlushko and B. Ilic. Tunable pinning in superconducting films with magnetic microloops. *Appl. Phys. Lett.* **89**, 182505–(3) (2006)

[Sil07] A. V. Silhanek, W. Gillijns, V. V. Moshchalkov, V. Metlushko, F. Gozzini, B. Ilic, W. C. Uhlig and J. Unguris. Manipulation of the vortex motion in nanostructured ferromagnetic/superconductor hybrids. *Appl. Phys. Lett.* **90**, 182501–(3) (2007)

[Smi86] D. P. E. Smith and G. Binnig. Ultrasmall scanning tunneling microscope for use in a liquid-helium storage dewar. *Rev. Sci. Instrum.* **57**, 2630–2631 (1986)

[Sol07] V. F. Solovyov and H. J. Wiesmann. Application of low-angle polishing for rapid assessment of the texture and morphology of thick film $Y_1Ba_2Cu_3O_7$ superconducting tapes. *Physica C* **467**, 186–191 (2007)

[Son00] J. E. Sonier, J. H. Brewer and R. F. Kiefl. μSR studies of the vortex state in type-II superconductors. *Rev. Mod. Phys.* **72**, 769–811 (2000)

[Spa07] M. Sparing, E. Backen, T. Freudenberg, R. Hühne, B. Rellinghaus, L. Schultz and B. Holzapfel. Artificial pinning centres in YBCO thin films induced by substrate decoration with gas-phase-prepared Y_2O_3 nanoparticles. *Supercond. Sci. Technol.* **20**, S239–S246 (2007)

[Str08] E. W. J. Straver, J. E. Hoffman, O. M. Auslaender, D. Rugar and K. A. Moler. Controlled manipulation of individual vortices in a superconductor. *Appl. Phys. Lett.* **93**, 172514–(3) (2008)

[The07] T. Thersleff. Private communication. Unpublished Work (2007)

Bibliography

[Tin96] M. Tinkham. *Introduction to superconductivity*. Dover publications, INC, Mineola, New York (1996)

[Ton02] A. Tonomura. Lorentz microscopy of vortices in superconductors. *Journal of Electron Microscopy* **51**, S3–S11 (2002)

[Vil07] J. E. Villegas, C.-P. Li and I. K. Schuller. Bistability in a superconducting Al thin film induced by arrays of Fe-nanodot magnetic vortices. *Phys. Rev. Lett.* **99**, 227001–(4) (2007)

[Vil08] J. E. Villegas, K. D. Smith, L. Huang, Y. Zhu, R. Morales and I. K. Schuller. Switchable collective pinning of flux quanta using magnetic vortex arrays: Experiments on square arrays of Co dots on thin superconducting films. *Phys. Rev. B* **77**, 134510–(5) (2008)

[Vol02] A. Volodin, K. Temst, C. Van Haesendonck, Y. Bruynseraede, M. I. Montero and I. K. Schuller. Magnetic-force microscopy of vortices in thin niobium films: Correlation between the vortex distribution and the thickness-dependent film morphology. *Europhysics Letters* **58**, 582–588 (2002)

[Vu93] L. N. Vu, M. S. Wistrom and D. J. Van Harlingen. Imaging of magnetic vortices in superconducting networks and clusters by scanning squid microscopy. *Appl. Phys. Lett.* **63**, 1693–1695 (1993)

[Wac02] A. Wachowiak, J. Wiebe, M. Bode, O. Pietzsch, M. Morgenstern and R. Wiesendanger. Direct observation of internal spin structure of magnetic vortex cores. *Science* **298**, 577–580 (2002)

[Wad92] A. Wadas and H. J. Hug. Models for the stray field from magnetic tips used in magnetic force microscopy. *J. Appl. Phys.* **72**, 203–206 (1992)

[Wad03] Y. Wada, K. Kuroda and T. Takami. Fabrication of YBCO multi-layer wiring with a polished PBCO insulator. *IEEE Trans. Appl. Supercond.* **13**, 817–820 (2003)

[Wan96] Z. Wang, A. Kawakami, Y. Uzawa and B. Komiyama. Superconducting properties and crystal structures of single-crystal niobium nitride thin films deposited at ambient substrate temperature. *J. Appl. Phys.* **79**, 7837–7842 (1996)

[Wei04] M. Weisheit, L. Schultz and S. Fähler. Textured growth of highly coercive $L1_0$ ordered FePt thin films on single crystalline and amorphous substrates. *J. Appl. Phys.* **95**, 7489–7491 (2004)

[Whi08] J. S. White, S. P. Brown, E. M. Forgan, M. Laver, C. J. Bowell, R. J. Lycett, D. Charalambous, V. Hinkov, A. Erb and J. Kohlbrecher. Observations of the configuration of the high-field vortex lattice in $YBa_2Cu_3O_7$: Dependence upon temperature and angle of applied field. *Phys. Rev. B* **78**, 174513–(13) (2008)

[Wie94] R. Wiesendanger. *Scanning probe microscopy and spectroscopy*. Cambridge university press (1994)

[Wu87] M. K. Wu, J. R. Ashburn, C. J. Torng, P. H. Hor, R. L. Meng, L. Gao, Z. J. Huang, Y. Q. Wang and C. W. Chu. Superconductivity at 93 K in a new mixed-phase Y-Ba-Cu-O compound system at ambient pressure. *Phys. Rev. Lett.* **58**, 908–910 (1987)

[Zhe92] X.-Y. Zheng, D. H. Lowndes, S. Zhu, J. D. Budai and R. J. Warmack. Early Stages of $YBa_2Cu_3O_{7-\delta}$ epitaxial-growth on MgO and $SrTiO_3$. *Phys. Rev. B* **45**, 7584–7587 (1992)

[Zhu03] A. A. Zhukov, P. A. J. de Groot, V. V. Metlushko and B. Ilic. Narrow commensurate states induced by a periodic array of nanoscale antidots in Nb superconductor. *Appl. Phys. Lett.* **83**, 4217–4219 (2003)

[Zur07] M. A. Zurbuchen, W. Tian, X. Q. Pan, D. Fong, S. K. Streiffer, M. E. Hawley, J. Lettieri, Y. Jia, G. Asayama, S. J. Fulk, D. J. Comstock, S. Knapp, A. H. Carim and D. G. Schlom. Morphology, structure, and nucleation of out-of-phase boundaries (OPBs) in epitaxial films of layered oxides. *Journal of Materials Research* **22**, 1439–1471 (2007)

List of Figures

2.1 Spatial variation of the microscopic magnetic field $h(x)$ and of the local density of superconducting electrons n_s at the interface between superconducting and normal state [Tin96]. 6

2.2 Sketch of the phase diagram of type-II superconductors [Buc04]. 7

2.3 (a) Profiles of magnetic field and order parameter for flux lines with lattice spacings of 4λ (solid lines) and 2λ (thin lines). The dashed line shows the magnetic field of an isolated vortex. (Calculations made using the GL theory by Brandt [Bra02a]). (b) Magnetic field lines above the surface (top) and profiles of order parameter and magnetic field (bottom) for a superconducting film with thickness $d = 8\lambda$ and vortex lattice spacing of 10λ calculated from GL theory [Bra09a]. 9

2.4 The magnetization curve for a $Nb_{55}Ta_{45}$ alloy (a) without defects (b) with a high amount of defects (1 kG=0.1 T). (Adapted from [Buc04]) 12

2.5 Bean's model approximation for the internal flux-density profiles in a thick slab for (a) increasing and (b) decreasing external field H_{ext}. H_s is the maximum field that can be completely screened [Tin96]. 13

3.1 (a) First local image of the Abrikosov lattice. Bitter decoration on Pb4-at%In rod at 1.1 K and 0.3 mT. Black points are decorated vortices [Ess67]. (b) STM image of the Abrikosov lattice in a Nb_3Sn single crystal at 1.8 K and 1 T [Hes89]. 16

3.2 Single-vortex resolution with different experimental techniques: range of magnetic induction versus scan-range [Fas08]. The abbreviations are: scanning tunnelling microscopy (STM), magnetic decoration (MD), Lorentz microscopy (LM), magnetic force microscopy (MFM), scanning SQUID microscopy (SSM), scanning Hall probe microscopy (SHPM), scanning electron microscopy (SEM) and magneto-optical imaging (MO). 1 G = 0.1 mT. 21

List of Figures

4.1 (a) Sketch of the MFM imaging procedure. The magnetized MFM tip scans above the surface of the sample at a given distance z. The feedback loop is not active. The resonance frequency shift is detected. (b) Van der Waals force. ... 24

4.2 (a) Sketch of the Cryogenic SFM (Omicron) microscope. Adapted from [Omicron user guide]. (b) Microscope head. ... 26

4.3 Interferometric detection scheme. The Light/Detector Unit (LDU) (Adapted from the Omicron user guide). The measured signal intensity varies with the distance d between the cantilever and the fiber cleaved edge. ... 27

5.1 Typical roughnesses of YBCO films measured by AFM. (a) CSD film, (b) *off-axis* PLD film. The peak-to-valley roughness of each film is shown by the scale bar. Scanning area is 4 μm × 4 μm. ... 30

5.2 LT-MFM images of vortices at 10 K frozen in different fields: (a) $H = -3$ mT, (b) $H = +2$ mT. The disordered vortex lattice points to a strong pinning by the natural defects in YBCO films. The scale bar represents the relative value of the measured signal: red color corresponds to a repulsive tip-vortex force and blue to an attractive one. Scanned area is 4 μm × 4 μm. (c) and (d) are 3D views of the individual vortices zoomed from (a) and (b) scans respectively. 32

5.3 The LT-MFM image of vortices at 30 K and -2 mT. ... 33

5.4 (a) AFM image of the *off-axis* PLD film with slightly increased roughness. (b) LT-MFM image on the same position, $H = 2$ mT, $T = 10$ K. White contours depict flux lines imaged as blue objects. The scanned area is 4 μm × 4 μm. ... 34

5.5 (a) Thermally activated drift of vortices, $T = 10$ K, $H = -3$ mT. The scanned area is 4 μm × 4 μm. (b) Zoom-in of the vortex elongated in the y direction during the scanning process. ... 35

5.6 (a) $U(I)$ curves measured on the structured YBCO bridge (3 mm long, 100 μm wide and 300 nm thick) at zero field and at different temperatures. The critical current is estimated to be 0.3 A at 10 K. (b) LT-MFM images on the bridge at +2 mT and 10 K. The flux lines are blue objects with a fuzzy shape. (c) LT-MFM scan on the same position after current pulses with an amplitude of 1 A have been passed through. Despite the fact that the current is largely above I_c, the flux distribution did not change. (d) Magneto-optical images of the whole bridge at 10 K and 7.4 mT. (f) The same sample measured at 17.2 mT in the middle of the bridge. (e) and (g) measured at 17.2 mT on the left and right defects. The field penetrates through the whole bridge. ... 37

List of Figures

5.7 (a) Sketched representation of all defects mentioned in this section. (b) Sketch shows the possible shift between the vortex visible on the surface and the nanodefect that pin it. Vortex is shown by the white line. 38

5.8 Room temperature AFM images (15 μm × 15 μm) of (a) $La_{0.7}Sr_{0.3}MnO_3$ pyramids on the STO substrate. (b) The same position after the deposition of a flat 200 nm thick YBCO film. (c) The same position after nanoscale polishing. The insets show the line profiles. 40

5.9 (a) CAD sketch of the search mask for the 5 mm×10 mm sample. The crosses are visible at the surface of the film during manual approach. (b) and (c) are optical microscope images at the positions marked by orange and blue color, respectively. (d) 30 μm × 30 μm AFM scan at the position marked with a black square in picture (b). (e) Line profile of the AFM scan. 42

6.1 10 μm × 10 μm AFM image of a PLD YBCO film (a) as-prepared and (b) after planar polishing. 47

6.2 10 μm × 10 μm AFM image of a PLD YBCO film (a) as prepared – 35 nm high precipitates are seen, (b) after planar polishing – roughness about 10 nm and (c) 4 μm × 4 μm LT-MFM image after polishing: -0.3 mT, 30 K, in-field cooling. 48

6.3 Sketch of a polished wedge with a reduction of film thickness from 200 to 50 nm on a 5 mm length. 49

6.4 High resolution SEM images along the polished wedge measured at a distance of (a) 3 mm, (b) 4 mm, (c) 5 mm, (d) 6 mm, (e) 7 mm and (f) 8 mm from the thick edge of the sample. 51

6.5 HR EBSD (a) on the as-prepared film and (b) on the thinnest part of the wedge. Black represents non-indexed regions. White color corresponds to the YBCO matrix, blue (small points) corresponds to the epitaxially grown 20–30 nm small $BaHfO_3$ precipitates, all other colors (larger points) belong to the different mismatch angle between $BaHfO_3$ particles and the YBCO matrix. . . 52

6.6 (a) (100) pole figure of the pure YBCO film. (b) (111) pole figure of $BaHfO_3$. One can clearly see the epitaxial relationship of the perovskite particles to the YBCO matrix. The non-cubic textured components in pole figure (b) come from randomly distributed clusters of precipitates. Pole figures were calculated from measured EBSD maps. 52

6.7 Cross-sectional FIB cut of (a) an as-grown and (b) a polished FePt/Nb double-layer. (c) Sketch of the achieved topology. 53

95

List of Figures

7.1 Vortex imaging on NbN film, field cooled in $-3\,\text{mT}$ at (a) $50\%T_c = 8\,\text{K}$, (b) $62\%T_c = 10\,\text{K}$ and (c) $75\%T_c = 12\,\text{K}$ (d) $87\%T_c = 14\,\text{K}$. Scanning area: $4\,\mu\text{m} \times 4\,\mu\text{m}$, scanning distance $z = 30\,\text{nm}$. 57

7.2 Temperature dependence of the vortex profile. Blue points are the measured signal, the solid line is a fit with the monopole-monopole model. 58

8.1 Sketch of a cross-sectional cut of the FM/SC hybrid structure and the MFM tip scanning above the surface. 64

8.2 Magnetic hysteresis loop, showing a magnetic vortex behavior with vortex nucleation and annihilation fields. Within the inner loop, between $-30\,\text{mT}$ and $+30\,\text{mT}$, the magnetization is reversible (vortex branch). The inset shows a SEM image of Py dots covered by a 100-nm thick Nb layer. The marked points on the loop correspond to the panels (a)-(e) in Fig. 8.3, respectively. . . 65

8.3 To reach the magnetic vortex state the in-plane field H_y was varied along the hysteresis loop starting from saturation at (a) $100\,\text{mT}$, (b) $25\,\text{mT}$, (c) 0, (d) through applying a negative field less than the magnetic-vortex annihilation field ($-25\,\text{mT}$) to the magnetic-vortex state at $0\,\text{mT}$ (e) and (e'). Color bars give the measured Δf signal. A small out-of-plane field of $10\,\text{mT}$ was permanently applied to set the orientation of the magnetic vortex (attractive interaction). The scanning area was $4\,\mu\text{m} \times 4\,\mu\text{m}$, the scanning distance 75 nm, $T = 14.6\,\text{K}$. The white circle represents the location of the Py dot. 66

8.4 (a) Line profiles taken from MFM images at different in-plane fields along the line going in the y direction through the center of the Py dot marked by the white circle (Fig. 8.3). The signal is proportional to the second derivative of the z component of the magnetic stray field of the Py dot. (b) Stray field distribution just above the surface of the Py dot in the vortex state. The estimated values are based on micromagnetic calculations using the LLC-program [Sch06]. 66

8.5 Visualization of superconducting vortices pinned by Py dots at $6.1\,\text{K}$ ($72\%T_c$). For a better visualization of the vortex positions the "background" image was subtracted. The frozen fields ΔH_z are: (a) $-0.5\,\text{mT}$, (b) $-1\,\text{mT}$ and (c) $-1.5\,\text{mT}$. Superconducting vortices are visualized as red spots. The white circles represent the locations of four Py dots. 68

8.6 Distribution of superconducting vortices in a Nb film in the presence of Py dots in the magnetic-vortex state measured at $T = 6.1\,\text{K}$ and $\Delta H_z = -1\,\text{mT}$. The superconducting vortices and the MFM tip exhibit an attractive interaction. The distance between the tip and the surface (h_2) decreases from panel (a), $h_2 = 90\,\text{nm}$, to panel (d), $h_2 = 60\,\text{nm}$. White circles represent the positions of the Py dots. The arrows mark the depinned vortices 70

List of Tables

4.1 Characteristic features of the MFM tip. 25

5.1 Critical current density (J_c) evaluated for a thin YBCO stripe by different methods 38

7.1 Temperature dependence of the tip-sample distance in the monopole-monopole model 59

7.2 Pinning force per vortex at different temperatures evaluated from the transport measurements for $\mu_0 H \approx 0$ 60

Acknowledgments

This chapter gives me the unique opportunity to express my gratitude to all those who supported me during my PhD. Without you, my dear colleagues and friends, this work would have never become as it is.

First of all I am deeply indebted to my *Doktorvater* Prof. Ludwig Schultz who inspired me from our first meeting and throughout the whole time of my work with his enthusiasm, his encouragement and his fascination for physics and material science. I would like to appreciate his open ears to all possible questions and problems, his support and his valuable hints and comments to my work. I am very thankful to him for giving me the opportunity to carry out my PhD work in the IFW Dresden where the pleasant working atmosphere is combined with unlimited opportunities of doing research and improving professional and soft skills. These unforgettable years of my PhD were not only important for me to develop as a scientist but also the experience I got in the IFW is indispensable for my future life.

I would like to gratefully and sincerely thank my direct supervisor Dr. Volker Neu who accompanied me through all the depths and heights of my work always having time to discuss and sharing my sorrows about our microscope and all problems that appeared with regard to research. During the long measurements days and nights, with seemingly unsolvable technical problems, with strange unexpectable behavior of the microscope – he never left me alone. Being always patient with my typos he commented and corrected all my scientific texts with great enthusiasm and deep interest. I would like to appreciate his encouraging suggestions and valuable ideas he had during many scientific discussions.

My second supervisor, Dr. Bernhard Holzapfel, I would like to thank deeply for his boundless optimism, for this restless good mood and of course for his valuable advises and many inspiring ideas that appeared during numerous discussions in the coffee-room or at the project meetings. I would like to acknowledge his tremendous contribution that gave me the opportunity to present my results at different international conferences and workshops.

I am very grateful to Prof. Vitali Metlushko for his interest in my work and his willingness to review my thesis. Moreover, I am thankful to him for the pleasant and fruitful cooperation where he supported me with the exciting samples – these measurements are an important part of my thesis. His inspiring ideas and rapid answers to my e-mails made me believe that Chicago is very close to Dresden.

As this thesis is dealing with the characterization of thin films, I am deeply thankful to the following people who prepared the samples or participated in the preparation and structuring processes.

I am very grateful to Elke Reich (Backen) who always listened to my possible and impossible wishes and prepared different kinds of PLD YBCO films (thin, thick, smooth, rough) supporting the quick and effective realization of all, sometimes quite "crazy" ideas.

Dr. Jens Hänisch I thank for valuable discussion in the topic of pinning. He and Karolin Tscharntke were supporting me with PLD YBCO films in the first year of my PhD.

Acknowledgments

Thomas D. Thersleff I would like to thank for the CSD films preparation and for the FIB cuts and FIB holes in YBCO films.
I am thankful to Dr. Sebastian Engel for CSD film preparation and for HR EBSD measurements on the wedge polished film.
Maria Sparing and the group of Prof. Xavier Obradors from ICMAB Barcelona I am thankful for the preparation of different kinds of artificial defects: Y_2O_3 nanoparticles, CeO_2 nanodots and nanowalls and $La_{0.7}Sr_{0.3}MnO_3$ pyramids. Dr. Günter Fuchs I would like to thank for the heavy-ion irradiation of the YBCO film.
Dr. Silvia Haindl and Jan Engelmann I am thankful for the preparation and J_c measurement on the NbN film.
I would like to thank Prof. Vitali Metlushko for the Py/Nb hybrid structures.
My great thank is going to Dr. Jens Ingolf Mönch and his team from the clean-room, Sandra Sieber and Barbara Eichler, for structuring the films in all possible and even impossible ways. I deeply appreciate their work, the days they spent and their interest to my work and enormous support during the creation and development of the searching mask.
Dr. Thomas Goedsche I am thankful for the many CAD images he prepared for me.
Dr. Nadezda Kozlova I would like to thank for introducing me to nano-painting (preparing tiny silver contacts on extremely small surfaces).
Martin Kidszun, Konrad Güth and Michael Kühnel I would like to thank for the gold deposition.
I gratefully appreciate Marina Gründlich's long-term experience in metallography, interest to my work, patience and important hints that contributed significantly to the successful development of the nanoscale wedge polishing procedure.
Dagmar Meier I would like to thank for her tremendous interest to the polishing procedure and productive cooperation.
The importance of technical support of Katja Berger, Martin Lange and Stefan Pofahl cannot be emphasized enough. Ulrike Fiedler I would like to thank for the inductive T_c measurements and Ulrike Besold for always being willing to share the most important liquid of my thesis - liquid helium.
Although the data is not included, I would like to appreciate the group of Prof. Paul Seidel, Dr. Frank Schmidl and Veit Große, from FSU Jena for the very interesting and perspective collaboration on the topic of polishing of YBCO films for SQUID applications.
Another very exciting collaboration, which also could not be included in the manuscript, was with Dr. Dmytro S. Inosov, MPI Stuttgart. The huge amount of the data we got on the new superconducting materials (iron-pnictides), he supported, during the last month of my PhD work brought not only new interesting results but also provoked me to better understand the nature of superconductivity. His enormous encouragement and intellectual curiosity inspired me to look deeply into different physical questions.

Dr. Ulrike Wolff, the mother of our SFM group, I deeply appreciate for her being always ready-to-help, for her deep interest to my work and open ears for all problems. Her mobile phone was never switched-off – she was always ready to listen to my problems regarding our microscope, answering my questions during the night, on vacation or during conferences. Firstly, introducing me to the mystery of low temperature magnetic force microscopy she continued to share the microscope-life with me during the whole period of my research strongly supporting me with all activities, with all unexpected complications, or sudden accidents that

Acknowledgments

our dear microscope prepared for us from time to time. Her steady hand very often assisted me with tiny works and was indispensable for the microscope-head repairs as well as for the MFM tip preparation.

For the strong support with magneto-optical measurements I would like to thank Dr. Jeffrey McCord, Dr. Rudolf Schäfer, Stefan Pofahl, Dr. Ryan Lai, Norbert Martin and Felix Kurth.

The project HIPERCHEM[1] and all its participants I would like to thank for the interesting and valuable cooperation on the topic of YBCO thin films.

I would like to thank Dr. Manfred Wolf for the micromagnetic calculation of the magnetic-vortex state and for his accurate reading and valuable comments of our joint publications.

My department leader Dr. Rudolf Schäfer I am deeply grateful for his strong support and encouragement and for his inspiring ideas and valuable advises to the magnetic part of my work. The friendly atmosphere he is spreading in the department was a very important prerequisite of the successful work.

According to this I would like to thank all my colleagues from department 24 for their help and support, for the nice working atmosphere and inspiring and exciting Friday-meetings.

Being on a junction of the topic I also spent a lot of time with department 26. Dear coffee-room group thank you all for the warm atmosphere and long fruitful discussions around the coffee-table. Here my special thanks is going to Dr. Ruben Hühne who supported me with writing the HIPERCHEM reports.

I would like to thank Dr. Filippo Giubileo and Prof. Anna Maria Cucolo for the unforgettable two weeks at the University of Salerno, Italy, where I was able to experience the challenges of tunneling microscopy.

Alexander Mietke I would like to thank for being the best summer student I could ever imagine.

I would like to say thank you to Katja Clausnitzer and Silja Schumann for the perfect office work they are doing that makes the IFW life even more pleasant.

I should not forget the support of the IFW workshop, facility management and IT department.

My LATEX-gurus, Dr. Marko Herrmann, Dr. Dmytro S. Inosov amd Dr. Carsten Weidner, my greatest thanks.

The PhD life is, of course not only measuring and writing. In every student life the very important part is communication. Thus, I would like to express my gratitude to our lunch-group: Elke Reich, Dr. Marko Herrmann, Maria Sparing, Jan Engelmann, Martin Kidszun, Dr. Silvia Haindl, Dr. Jens Hänisch, Thomas D. Thersleff, Dr. Kazumasa Iida, David Geißler and Dr. Thomas G. Woodcock for the pleasant time we spent in the IFW and outside. Thank you all, you have been a great encouragement for me during these years.

Our SFM group: Dr. Volker Neu, Dr. Ulrike Wolff, Cristina Bran, Silvia Vock, Marietta Seifert, Dr. Ajit Patra, Felix Fleischhauer, Dagmar Meier, Andreas Reisner, Katja Berger and Martin Lange I thank for the support, nice working atmosphere and unforgettable hiking tours.

Here, I would like to thank once again my friend, my best roommate, my colleague Elke Reich (Backen) for everything she did for me. Always ready to share my problems, to listen

---

[1] The reported results were obtained with research funding from the European Community under the Sixth Framework Programme Contract Number 516858: HIPERCHEM.

Acknowledgments

and to give important hints she created the greatest office atmosphere I can imagine. She thoroughly read my English and German writing trying to explain why the language needs articles and how I should use them.

My friend and colleague Dr. Silvia Haindl I thank for being always critical to my work and of my writing, for long discussions on different scientific topics, for her unlimited ready-to-help and her clear head full of "crazy" scientific ideas.

No words can really express the importance of the allround support of my dear husband Jochen Werner. His infinite patience, his incredible optimism, his love and tolerance as well as his ability to listen and to solve problems, always having valuable hints and ideas ready for me, were the indispensable foundation during all the years. My dear parents, Alexey and Lyudmila, I would like to owe my special thanks for their enormous support and absolute faith in me all my life long. They were always on my side and trusted me in all my decisions allowing and supporting me on my way.

I want morebooks!

Buy your books fast and straightforward online - at one of world's fastest growing online book stores! Environmentally sound due to Print-on-Demand technologies.

Buy your books online at
www.morebooks.shop

Kaufen Sie Ihre Bücher schnell und unkompliziert online – auf einer der am schnellsten wachsenden Buchhandelsplattformen weltweit! Dank Print-On-Demand umwelt- und ressourcenschonend produziert.

Bücher schneller online kaufen
www.morebooks.shop

KS OmniScriptum Publishing
Brivibas gatve 197
LV-1039 Riga, Latvia
Telefax: +371 686 204 55

info@omniscriptum.com
www.omniscriptum.com

Printed by Books on Demand GmbH, Norderstedt / Germany